AF352828

Nutrition and Women Bone Health

Nutrition and Women Bone Health

Editor

Jose M. Moran

MDPI • Basel • Beijing • Wuhan • Barcelona • Belgrade • Manchester • Tokyo • Cluj • Tianjin

Editor
Jose M. Moran
Metabolic Bone Diseases
Research Group,
University of Extremadura,
10003 Cáceres, Spain

Editorial Office
MDPI
St. Alban-Anlage 66
4052 Basel, Switzerland

This is a reprint of articles from the Special Issue published online in the open access journal *Nutrients* (ISSN 2072-6643) (available at: https://www.mdpi.com/journal/nutrients/special_issues/women_bone_health).

For citation purposes, cite each article independently as indicated on the article page online and as indicated below:

LastName, A.A.; LastName, B.B.; LastName, C.C. Article Title. *Journal Name* **Year**, *Volume Number*, Page Range.

ISBN 978-3-0365-4465-6 (Hbk)
ISBN 978-3-0365-4466-3 (PDF)

Contents

Editorial

Nutrition and Women's Bone Health

Jose M. Moran

Metabolic Bone Diseases Research Group, Nursing and Occupational Therapy College,
University of Extremadura, Avd. Universidad s/n, 10003 Caceres, Spain; jmmorang@unex.es

Citation: Moran, J.M. Nutrition and Women's Bone Health. *Nutrients* **2022**, *14*, 763. https://doi.org/10.3390/nu14040763

Received: 21 January 2022
Accepted: 10 February 2022
Published: 11 February 2022

Publisher's Note: MDPI stays neutral with regard to jurisdictional claims in published maps and institutional affiliations.

Nutrition is a key element that has the potential to reduce bone loss or fracture risk. This Special Issue presents some recent promising developments in nutrition and women's bone health in the form of four original contributions and one systematic review. Overall, key aspects of the interaction between different nutrients and women's bone health are presented through research designs of the highest quality, covering both randomized clinical trials and observational studies and headed in all cases by leading scientists in the areas of nutritional and clinical research.

By implementing a randomized, double-blind, placebo-controlled clinical trial using intention-to-treat analysis, Vallibhakara and colleagues addressed the effect of nutritional vitamin E supplementation in osteopenic postmenopausal women. This study addressed bone health by measuring changes in bone turnover markers after supplementation for 12 weeks. Most of the study participants showed good nutritional status (mean BMI 22 kg/m^2) and high adherence was reported in both groups [1]. Those in the study group received 400 IU of mixed tocopherol on a daily basis for 12 weeks. Tablets contained 20% delta-tocopherol, 1% beta-tocopherol, 62% gamma-tocopherol and 10% alpha-tocopherol (Nat E®, Mega Lifesciences Public Company Limited, Samutprakarn, Thailand). While a comparison of the bone turnover markers, CTX and PINP, showed no significant difference between the vitamin E and placebo arms both at baseline and after 12 weeks of supplementation, the mean difference in the bone resorption marker, CTX, from baseline to 12 weeks was significantly different between the vitamin E group and the placebo group (-0.003 ± 0.09 and 0.121 ± 0.15, respectively ($p < 0.001$). As the bone formation markers in both groups were not significantly changed between groups in this study, Vallibhakara and colleagues suggest that the beneficial effect of vitamin E supplementation on bone health in postmenopausal women may be targeted towards slowing the increase in the bone resorption marker (CTX), which may represent the mitigation of bone loss through antiresorptive activity.

Moschonis and colleagues, following up on their previous studies on the consumption of reduced-fat Gouda cheese fortified with vitamin D3 and its efficacy in reducing the prevalence of winter vitamin D deficiency in a population of postmenopausal women in Greece [2], studied the effects of vitamin D-enriched cheese on serum PTH concentrations and selected biomarkers of bone remodeling in early or late postmenopausal women with adequate or insufficient vitamin D at baseline. Their interesting approach reveals the usefulness of supplementation of traditional dietary foods and their potential benefit for female bone metabolism [3]. A randomized, controlled, single-blinded (i.e., blinded to study participants only), controlled dietary intervention study was designed to evaluate the effect of Gouda-type cheese fortified with vitamin D3 on serum concentrations of certain calciotropic hormones (i.e., 25(OH)D, PTH), bone formation (i.e., OC, P1NP), and bone resorption markers (i.e., TRAP-5b) in postmenopausal women. The intervention proposed using vitamin D-enriched Gouda cheese significantly increased serum 25(OH)D concentrations, prevented PTH increase and reduced bone resorption in early postmenopausal women with vitamin D insufficiency. Interestingly, the reduction in bone resorption reported in women with vitamin D insufficiency coincides with increased

25(OH)D concentrations, leading the authors to propose that their results may reflect a promising nutrient-based approach to enhancing vitamin D status in these women, and to bring positive changes in bone metabolism that may be protective toward the bone loss that follows menopause.

Fatty acids are key nutrients for health, and several studies have reported an association between bone mineral density (BMD) and fatty acid intake. Robust evidence from observational studies has reported how total PUFA intake, especially n-3 and n-6 PUFA, improves BMD and even reduces the risk of fracture [4]. From an observational standpoint, Roncero-Martin and colleagues [5] evaluated the associations between serum levels of different PUFAs (n-6 and n-3), MUFAs and SFAs with bone density determined by quantitative bone ultrasound (QUS), peripheral quantitative computed tomography (pQCT) and dual-energy X-ray absorptiometry (DXA) in a sample of Spanish postmenopausal women. Independent risk factors for low BMD (T-score ≤ 1) were derived using logistic regression analysis. Higher BMI (OR = 0.893; 95% CI 0.841–0.948), $p < 0.001$) and higher plasma n-3 PUFA (OR = 0.751; 95% CI 0.587–0.960, $p = 0.022$) were identified as protective factors against low bone mass. This study reports a strong statistically independent and positive association between BMD and plasma n-3 PUFA levels, demonstrating the physiological and biochemical relevance of total plasma omega-3 fatty acids.

Based on the hypothesis that dietary total antioxidant capacity is positively associated with bone mass and negatively associated with the risk of osteoporosis in premenopausal and postmenopausal women, and based on a study of data obtained from the Korea National Health and Nutrition Examination Survey 2008–2011, Kim and colleagues [6] present data obtained from a total of 8230 female participants. The total antioxidant capacity of the diet (TAC) is a useful parameter to evaluate the overall antioxidant capacity of foods; instead of a simple addition of individual dietary antioxidants, the dietary TAC yields comprehensive information on the cumulative antioxidant capacities of different diets and may be helpful in the assessment of the preventive effects of antioxidants in different pathologies. Dietary TAC was associated positively with bone mass at the lumbar spine and femoral neck and inversely associated with the risk of osteoporosis in this cohort of postmenopausal Korean women, while in the group of premenopausal women, dietary TAC was positively associated with the bone mineral content of the lumbar spine and total femur. The results presented by Kim and colleagues in this Special Issue suggest, from an observational perspective, that consumption of high TAC foods in the diet, such as grapes, radish leaves, bell pepper paste, oranges and spinach, may improve bone health, especially in postmenopausal women, and they open the door to exciting research from the perspective of the association between the consumption of these foods and the risk of osteoporosis from an experimental point of view.

The novel advances presented in this Special Issue are further complemented by a comprehensive review of the scientific literature that offers thorough background on plant-derived compounds that potentially could be used to support bone health in perimenopausal and postmenopausal women. The review covers compounds with antiosteporotic characteristics, covering both in vitro and in vivo studies as well as clinical trials. A detailed biochemical and clinical review of the potential effects of phytoestrogens and other botanicals is provided. Słupski and colleagues [7] identify these botanicals studied as a major source of bioactive compounds, many waiting to be further investigated for their possible beneficial effects on bone health and in particular for the treatment and prevention of osteoporosis.

The current Special Issue presents progress on the topic of women's bone health and nutrition, showing the important role that this interaction plays in human health and in different populations. The advances shown are of great interest from a clinical perspective, and these results are the starting point of what should be the basis for future research, with a growing presence of experimental studies based on different populations and ages, allowing for a deeper understanding of the ultimate interaction that takes place between nutrition and bone health in women.

Funding: This research received no external funding.

Acknowledgments: J.M.M. would like to thank all of the contributing authors, to our Special Issue "Nutrition and Women Bone Health", all the reviewers for their work in evaluating the submitted articles, and the editorial staff of *Nutrients*.

Conflicts of Interest: The authors declare no conflict of interest.

References

1. Vallibhakara, S.A.-O.; Nakpalat, K.; Sophonsritsuk, A.; Tantitham, C.; Vallibhakara, O. Effect of Vitamin E Supplement on Bone Turnover Markers in Postmenopausal Osteopenic Women: A Double-Blind, Randomized, Placebo-Controlled Trial. *Nutrients* **2021**, *13*, 4226. [CrossRef] [PubMed]
2. Manios, Y.; Moschonis, G.; Mavrogianni, C.; van den Heuvel, E.; Singh-Povel, C.M.; Kiely, M.; Cashman, K.D. Reduced-fat Gouda-type cheese enriched with vitamin D3 effectively prevents vitamin D deficiency during winter months in postmenopausal women in Greece. *Eur. J. Nutr.* **2017**, *56*, 2367–2377. [CrossRef] [PubMed]
3. Moschonis, G.; van den Heuvel, E.G.; Mavrogianni, C.; Manios, Y. Effect of Vitamin D-Enriched Gouda-Type Cheese Consumption on Biochemical Markers of Bone Metabolism in Postmenopausal Women in Greece. *Nutrients* **2021**, *13*, 2985. [CrossRef] [PubMed]
4. Lavado-Garcia, J.M.; Roncero-Martin, R.; Moran, J.M.; Pedrera-Canal, M.; Aliaga, I.; Leal-Hernández, O.; Martin, S.R.; Canal-Macias, M.L. Long-chain omega-3 polyunsaturated fatty acid dietary intake is positively associated with bone mineral density in normal and osteopenic Spanish women. *PLoS ONE* **2018**, *13*, e0190539. [CrossRef] [PubMed]
5. Roncero-Martín, R.; Aliaga, I.; Moran, J.; Puerto-Parejo, L.; Rey-Sánchez, P.; de la Luz Canal-Macías, M.; Sánchez-Fernández, A.; Pedrera-Zamorano, J.; López-Espuela, F.; Vera, V.; et al. Plasma Fatty Acids and Quantitative Ultrasound, DXA and pQCT Derived Parameters in Postmenopausal Spanish Women. *Nutrients* **2021**, *13*, 1454. [CrossRef] [PubMed]
6. Kim, D.; Han, A.; Park, Y. Association of Dietary Total Antioxidant Capacity with Bone Mass and Osteoporosis Risk in Korean Women: Analysis of the Korea National Health and Nutrition Examination Survey 2008–2011. *Nutrients* **2021**, *13*, 1149. [CrossRef] [PubMed]
7. Słupski, W.; Jawień, P.; Nowak, B. Botanicals in Postmenopausal Osteoporosis. *Nutrients* **2021**, *13*, 1609. [CrossRef] [PubMed]

Article

Effect of Vitamin E Supplement on Bone Turnover Markers in Postmenopausal Osteopenic Women: A Double-Blind, Randomized, Placebo-Controlled Trial

Sakda Arj-Ong Vallibhakara [1,2], Katanyuta Nakpalat [3], Areepan Sophonsritsuk [4], Chananya Tantitham [4] and Orawin Vallibhakara [4,*]

[1] Faculty of Medicine, Bangkokthonburi University, Bangkok 10170, Thailand; dr.sakda@gmail.com
[2] Child Safety Promotion and Injury Prevention Research Center, Faculty of Medicine, Ramathibodi Hospital, Mahidol University, Bangkok 10400, Thailand
[3] Woman Health Centre, Chulabhorn Hospital, HRH Princess Chulabhorn College of Medical Science, Chulabhorn Royal Academy, Bangkok 10210, Thailand; Katanyuta_ne@hotmail.com
[4] Reproductive, Endocrinology & Infertility Unit, Department of Obstetrics and Gynaecology, Faculty of Medicine Ramathibodi Hospital, Mahidol University, Bangkok 10400, Thailand; areepan.sop@mahidol.ac.th (A.S.); ploy.dango@gmail.com (C.T.)
* Correspondence: orawinra38@gmail.com

Citation: Vallibhakara, S.A.-O.; Nakpalat, K.; Sophonsritsuk, A.; Tantitham, C.; Vallibhakara, O. Effect of Vitamin E Supplement on Bone Turnover Markers in Postmenopausal Osteopenic Women: A Double-Blind, Randomized, Placebo-Controlled Trial. *Nutrients* **2021**, *13*, 4226. https://doi.org/10.3390/nu13124226

Academic Editor: Jose M. Moran

Received: 31 October 2021
Accepted: 22 November 2021
Published: 25 November 2021

Publisher's Note: MDPI stays neutral with regard to jurisdictional claims in published maps and institutional affiliations.

Abstract: Vitamin E is a strong anti-oxidative stress agent that affects the bone remodeling process. This study evaluates the effect of mixed-tocopherol supplements on bone remodeling in postmenopausal osteopenic women. A double-blinded, randomized, placebo-controlled trial study was designed to measure the effect of mixed-tocopherol on the bone turnover marker after 12 weeks of supplementation. All 52 osteopenic postmenopausal women were enrolled and allocated into two groups. The intervention group received mixed-tocopherol 400 IU/day, while the control group received placebo tablets. Fifty-two participants completed 12 weeks of follow-up. Under an intention-to-treat analysis, vitamin E produced a significant difference in the mean bone resorption marker (serum C-terminal telopeptide of type I collagen (CTX)) compared with the placebo group (-0.003 ± 0.09 and 0.121 ± 0.15, respectively ($p < 0.001$)). In the placebo group, the CTX had increased by 35.3% at 12 weeks of supplementation versus baseline ($p < 0.001$), while, in the vitamin E group, there was no significant change of bone resorption marker ($p < 0.898$). In conclusion, vitamin E (mixed-tocopherol) supplementation in postmenopausal osteopenic women may have a preventive effect on bone loss through anti-resorptive activity.

Keywords: vitamin E; bone turnover; osteopenia; postmenopausal women; tocopherol; bone marker turnover; bone health

1. Introduction

Bone is a dynamic tissue continuously remodeling itself throughout life for bone homeostasis [1]. The remodeling process is based on the balanced activities of bone resorption and formation. This process requires communication between different bone cells, namely, cells of the osteoblast lineage (osteoblasts and osteocytes) and bone-resorbing cells (osteoclasts), which are organized in specialized units called bone multicellular units (BMU). The bone remodeling controlled by various local and systemic factors in different pathways, such as calcitonin, parathyroid hormone, and estrogen pathways, is the major hormonal regulator of osteoclastic bone resorption. In addition, cytokines, such as Interleukin-1 (IL-1), Interleukin-6 (IL-6), prostaglandin E2 (PGE2), and tumor necrotic factor α (TNF-α), play roles in the regulation of physiological bone remodeling. The major pathway of bone remodeling is controlled by the receptor activator of nuclear factor kappa-B (RANK), receptor activator of nuclear factor kappa-B ligand (RANKL), osteoprotegerin (OPG), canonical Wnt/β-catenin, and oxidative stress signaling pathways [2–5].

The oxidative stress pathway activates the differentiation of pre-osteoclast into osteoclast and provokes bone resorption. An increasing level of reactive oxygen species (ROS) results in rapid bone loss, especially in postmenopausal women [6].

A postmenopausal woman means a woman with an absence of menstruation of at least 12 months, resulting from the cessation of ovarian function [7]. The resultant reduction of the female reproductive hormone associated with aging process, such as osteoporosis, atherosclerosis, and cancer. Oxidative stress is the pathophysiological driver of the aging process and a cause of osteoporosis. Oxidative stress aggravates the osteoclastic activity in the bone remodeling process, without effect on the osteoblast. Moreover, bone resorption predominates bone formation, resulting in diminished bone density. As mentioned, these mechanisms/factors are part of the cause of osteopenia and osteoporosis in postmenopausal women and the elderly [8,9]. The same as seen in rheumatoid arthritis patients who have a high level of ROS that causes low bone mass even during reproductive age [10,11].

Osteopenia, or low bone mass, as defined by the World Health Organization, is when bone mineral density (BMD) value, reported as a T-score and measured by a dual-energy X-ray absorptiometry (DXA), is between T-score-1 and -2.5 [12]. The prevalence of osteopenia is high worldwide, which an estimated 43.9%, 40.3%, and 59.4% of postmenopausal women in the United States, India, and Thailand, respectively [13–15]. The relative risk for fracture is three times greater for osteopenic women compared to women having normal BMD. Moreover, Pasco et al. found that the osteoporotic fractures mostly come from osteopenic women, who represent as high as approximately 50% of the total fracture population [16]. The latest survey study in Thailand reveals morphometric vertebral fracture at 29% among postmenopausal women with osteopenia [17]. Although women with osteoporosis are at the highest risk for fracture, these high-risk individuals comprise less than 20% of all women who fracture. Therefore, strengthening bone quality and fall prevention in postmenopausal women with osteopenia, which is not indicated for pharmacologic treatment, are essential for minimizing morbidity and mortality from fractures. According to the American Association of Clinical Endocrinologists/American College of Endocrinology clinical practice guideline for postmenopausal osteoporosis treatment. There is a strong indication for pharmacologic therapy when patient has: (1) low bone mass and a history of a fragility fracture of hip or spine; (2) osteoporosis, as diagnosed by BMD measurement with a T-score of -2.5 or lower in the spine, femoral neck or distal radius; and (3) osteopenia, as diagnosed by BMD measurement with the fracture risk assessment tool (FRAX®) 10-year probability for major osteoporosis fracture being $\geq$20% or the 10-year probability of hip fracture being $\geq$3%. The first-line medical management for osteoporotic fracture prevention is bisphosphonate agents, including alendronate, risedronate and zoledronate, and the receptor activator of nuclear factor kappa-B ligand (RANKL) inhibitors, or denosumab [18]. Those medications inhibit osteoclast's function, which affects the risk of the bone freezing condition, rare but serious complications include osteonecrosis of the jaw (ONJ), and atypical femoral fracture (AFF). The ONJ is the presence of exposed bone in the mouth that fails to heal after appropriate intervention over a period of 6 or 8 weeks. Incidence has been reported between 0.001–0.06% depending on the dose and duration of anti-resorptive administration [19,20]. In contrast, AFF is a transverse fracture of the femoral shaft (diaphysis), defined by clinical criteria and radiographic appearance. The incidence ranged from 3.0 to 9.8 cases per 100,000 patient per year [21,22]. Drug holiday guidelines recommend discontinuing bisphosphonates after five years of treatment, depending on patient fracture risk to prevent those serious side effects.

Vitamin E, one of the lipid-soluble vitamins, is a well-known strong anti-oxidative agent or antioxidant. This consists of two isoforms: tocopherol (TF) and tocotrienol (T3). Each isoform divides into four distinct analogs, namely alpha (α), beta (β), gamma (γ), and delta (δ). Tocopherol is the most commonly available form of vitamin E supplement on the market. The major properties of vitamin E are reducing the reactive oxygen species (ROS), and numerous pro-inflammatory cytokines, such as IL-1, IL-6, PGE2, and TNF-α, which are in part responsible for the activation of osteoclastic activity [23–28]. From observational studies in postmenopausal women, low serum α-tocopherol and low intake of vitamin E in diet were associated with osteoporosis and hip fracture [29,30]. In the animal model study, Muhammad et al. (2012) found that tocopherol and tocotrienol can decelerate bone resorption activity and decrease bone loss [31–33].

The study aimed to evaluate the effect of vitamin E supplements in postmenopausal osteopenic women on bone by measuring bone turnover marker changes after a mixed-tocopherol supplement for 12 weeks. The investigation was based on the hypothesis that vitamin E could decrease the differentiation and activity of osteoclast that causes bone loss attenuation.

2. Materials and Methods

2.1. Study Design

The study protocol conformed to the principles of the Declaration of Helsinki. It was approved by the Ethical Clearance Committee on Human Rights Related to Research Involving Human Subjects Faculty of Medicine Ramathibodi Hospital, Mahidol University (MURA 2018/02) and Grants. The study protocol was also submitted to the Thai Clinical Trials Registry; TCTR (www.thaiclinicaltrials.org) Clinical trial registration no. TCTR20180419001/18-04-2018. The double-blinded, placebo-controlled randomized trial was undertaken to study the effect of mixed-tocopherol on changes in bone turnover marker (serum C-terminal telopeptide of type I collagen (CTX), a bone resorption marker, and serum N-terminal propeptide of type I procollagen (P1NP), a bone formation marker) after 12 weeks of supplementation. The primary outcome was (1) a change of bone resorption marker (CTX) and (2) bone formation marker (P1NP). The secondary outcome was any adverse events.

2.2. Participants

Postmenopausal osteopenic women were enrolled in the study between 1 May 2018 and 31 May 2019 at Menopause clinic, Department of Obstetrics and Gynecology, Faculty of Medicine, Ramathibodi Hospital, Bangkok. The informed and consent of study was obtained from all participants. Postmenopausal women with absent menstruation for at least one year, age over 45-year-old, osteopenia indicated by a bone mineral density measured by Dual-energy X-ray Absorptiometry T-score between −1 and −2.5 at the lumbar spine, total hip or femoral neck, and signed agreements to participate in the study were included. Exclusion criteria were serum 25-hydroxyvitamin D (serum 25(OH)D level) < 20 ng/mL, having a history of metabolic bone disease, having a history of cancer, having an eating disorder or malabsorption, taking a medication that affects bone metabolism, taking an anticoagulant, having taken a vitamin E supplement within 3 months, and unwillingness to accept the randomization. A pilot study was performed to estimate sample size from 16 participants. The means and standard deviation (SD) of C-terminal telopeptide of type I collagen (CTX) from both groups were measured and calculated by N4studies Application (Version 1.4.1). The mean in a treatment group (mixed-tocopherol supplement) was 0.35, and SD was 0.15. The mean in a control group (Placebo) was 0.55, and SD was 0.29. The ratio (control/treatment) = 1.00. Alpha (α) = 0.05, Z (0.975) = 1.959964, and Beta (β) = 0.20, Z (0.800) = 0.841621. Then, the calculated sample size was 21 participants in the treatment and an equal number in the control group. A 20% larger sample was used to accommodate for potential data loss. Finally, there were 26 participants in each group, or a total of 52 participants in this study.

2.3. Randomization, Blinding, and Intervention Protocol

All participants were randomly assigned in a 1:1 ratio into two groups by the computerized block of four randomizations. The participants and investigators were blinded to the group allocation. Only the pharmacist, who prepared these study tablets, knew the allocation. Participants in the study group received 400 IU of mixed-tocopherol one time per day for 12 weeks. Each tablet contained 20% delta-tocopherol, 1% beta tocopherol, 62% gamma-tocopherol, and 10% alpha tocopherol (Nat E$^®$, Mega Lifesciences Public Company Limited, Samutprakarn, Thailand). The dose of vitamin E was calculated based on a conversion from rat to human by body surface area [34,35]. Participants in the placebo group received the placebo tablets, which had a similar external appearance to the mixed-tocopherol tablets but contained soybean oil. The participants in both groups received 600 mg of calcium carbonate twice per day (total of 1200 mg per day) and 20,000 IU of vitamin D2 one time per week.

2.4. Data Collection and Measurements

At the time of enrollment, baseline characteristics, including age, parity, menopause age, time since menopause, medical history, current medication, prior calcium or vitamin D or vitamin E supplement, smoking history, drinking history, and exercise habits, were gathered. Physical examinations were performed, and the clinical data, including weight, height, body mass index, blood pressure, and any signs of bleeding tendency, were collected. For exploring the secondary cause of bone loss, baseline laboratory analysis of serum 25(OH)D level, creatinine, aspartate aminotransferase (AST), alanine aminotransferase (ALT), parathyroid hormone level (PTH), thyroid-stimulating hormone (TSH), and complete blood count (CBC) were performed.

The change of bone turnover markers (BTMs) represented the bone remodeling process and was the primary outcome of this study. The BTMs are byproducts of the bone remodeling process. They include formative bone marker; bone-specific alkaline phosphatase (B-ALP), procollagen type 1 N-terminal propeptide (P1NP), osteocalcin (O), and procollagen type 1 C-terminal propeptide (P1CP), and bone resorption marker; hydroxyproline (HYP), pyridinoline, tartrate-resistant acid phosphatase 5b (TRAP5b), deoxypyridinoline (DPD), carboxy-terminal cross-linked telopeptide of type 1 collagen (CTX-1), and amino-terminal cross-linked telopeptide of type 1 collagen (NTX-1). Those markers are used in clinical trials since they rapidly change, compared to BMD, and are easily collected from urine or serum. Currently, BTMs have already proved to be of clinical use as adjuvant tools for fragility fracture risk stratification and monitoring the treatment response and adherence monitoring. The bone remodeling process consists of the bone resorption phase, in which the main action is by osteoclasts catabolizing old bone (lasts for 4–6 weeks) and the subsequent bone formation phase in which the osteoblast secretes molecules that fill in cavities with osteoid, a connective tissue rich in collagen. This entire bone formation process lasts 4–5 months [36]. According to the International Osteoporosis Foundation and the International Federation of Clinical Chemistry and Laboratory Medicine Working Group on Bone Marker Standards, serum C-terminal telopeptide (CTX) and serum N-terminal propeptide of type I procollagen (P1NP) have been recommended as the standard marker of bone resorption and formative bone marker, respectively [37]. Serum C-terminal telopeptide (CTX) is a product of the breakdown of type I collagen, the major bone protein component. After the release of the osteoclast enzyme, bone proteins are digested and released as their fragments, e.g., N-telopeptide of type I collagen (NTX-I) or CTX-I, into serum, plasma, and urine. Serum CTX is a bone-specific marker and decreases during anti-resorptive treatment. In contrast, serum N-terminal propeptide of type I procollagen (P1NP) is derived from post-translational cleavage of type I procollagen during the bone formative phase. Procollagen type I enzymatically cleaves the N-propeptides and C-propeptides and subsequently becomes collagen type I, which is deposited in a quarter stagger array held together by the pyridinium cross-links: deoxypyridinoline and pyridinoline. Osteoblasts express their highest collagen concentration during the proliferative phase, bone alkaline phosphatase, which occurs during matrix maturation, and osteocalcin during mineralization. Serum P1NP is a bone-specific marker that weakens circadian variation and increases during bone formation-stimulating therapy [38].

In this study, both bone turnover markers were measured by automated electrochemiluminescence immunoassay (ECLIA) for CTX and PINP, with cobas E602 (Roche Diagnostic, Germany). The intra- and inter assays coefficient of variation were 1.6% (0.004 ng/mL) and 2.2% (0.006 ng/mL), for the serum CTX and 1.7% (0.524 ng/mL) and 2.6% (0.797 ng/mL) for PINP, respectively. The venous blood collection was performed in a fasting sample at 7–10 a.m. to minimize the diurnal variation on the day of enrollment and at the end of the 12th week of follow-up. The serum was collected using standard sampling with EDTA tubes.

Participants received a reminder phone call one week before the appointment. At the 12th week of follow-up, all participants visited with the allocation bottles and returned them to the investigators to count the remaining tablets. Participants who took at least 80 percent of calcium, vitamin D2, and the study tablets of all prescriptions were defined as having good compliance. The serum AST and ALT were monitored. All adverse effects were recorded by either interviewing by investigators or self-reporting by the participants.

2.5. Statistical Analysis

Statistical analysis was performed by STATA Version 15.0 (College Station, TX, USA). The results were analyzed by the intention-to-treat statistical method. The baseline characteristics and the laboratory data were analyzed by the Shapiro–Wilk test for normal or non-normal distribution. All quantitative variables were tested for normal distributions. The student *t*-test was used for the comparison of the continuous variables in parametric data. The Mann–Whitney U test was used for the comparison of continuous variables in nonparametric data. In all instances, two-tailed tests of statistical significance were used. For comparison of the bone turnover marker between points in time within the group, a paired *t*-test was used. Data were presented as mean $\pm$ standard deviation (SD) and median (minimum and maximum) if the data had a non-normal distribution. The intention to treat analysis was done, and the level of statistical significance was set to $p < 0.05$.

3. Results

3.1. Study Participants

A total of 52 participants were recruited. All the participants completed the study in the 12th week. All of the patients in the placebo group had good compliance with the assigned tablet, including the calcium and vitamin D2 supplement, based on the counting of the remaining tablets at 12 weeks. Only one in the vitamin E group had poor compliance (compliance rate 30 percent) with no difference in outcome versus the placebo group. A protocol flow chart is shown in Figure 1.

Table 1 shows the baseline characteristics of the participants in both groups. There were no significant differences in baseline characteristics.

3.2. Bone Turnover Marker

A comparison of the bone turnover markers, CTX and PINP, showed no significant difference between the vitamin E and placebo group at baseline and after 12 weeks of supplementation. However, the mean difference in the bone resorption marker, CTX, at baseline versus 12 weeks were significantly different between the vitamin E group and the placebo group (-0.003 ± 0.09 and 0.121 ± 0.15, respectively ($p < 0.001$)), and no difference was shown in the mean difference in the formative bone marker, PINP between the vitamin E group and placebo group (-0.25 ± 14.02 and 2.4 ± 10.88, respectively ($p = 0.430$)) as shown in Table 2. In addition, the placebo group was showed a significant increase in bone resorption marker, CTX, during the time period, up by 35.3% at 12 weeks compared with baseline ($p < 0.001$). Meanwhile, the vitamin E group showed no significant change of bone resorption marker at 12 weeks compared with baseline ($p < 0.898$), as shown in Figure 2. The bone formation markers in both groups were not significantly changed.

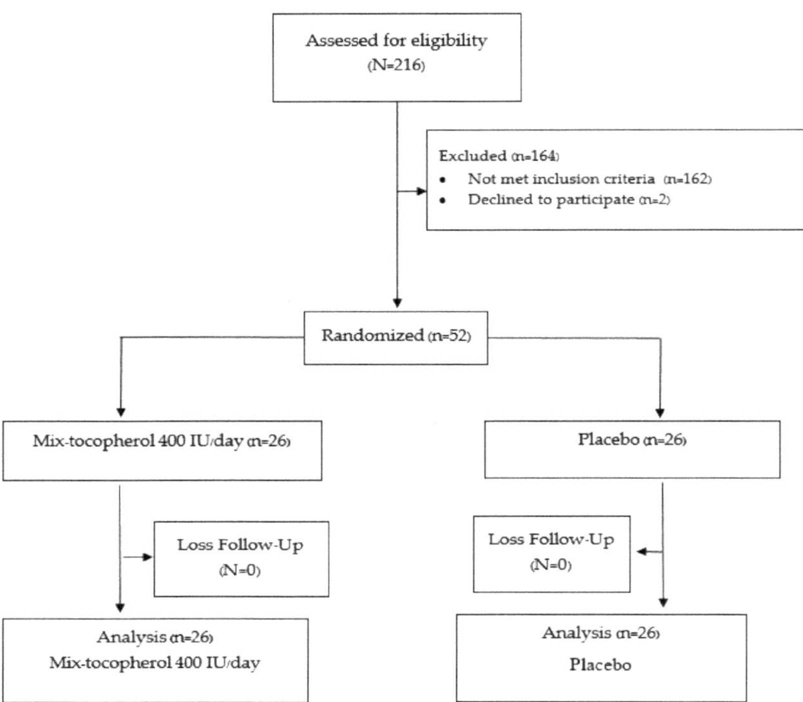

Figure 1. Protocol flow diagram according to CONSORT flow diagram 2010.

Table 1. Baseline characteristics of the participants.

Characteristics	Vitamin E ($n = 26$)	Placebo ($n = 26$)	p-Value
Age (years) [a]	63.15 ± 7.96	62.31 ± 6.50	0.65
Body mass index (kg/m^2) [a]	21.93 ± 3.61	22.56 ± 2.73	0.48
Age at menopause (years) [a]	48.77 ± 4.12	48.58 ± 4.16	0.87
Duration of postmenopausal period (years) [a]	13.54 ± 7.46	14.58 ± 7.80	0.63
Exercise [b]			0.58
No exercise	13 (50%)	15 (57.7%)	
Exercise (<150 min/week)	13 (50%)	11 (42.3%)	
Baseline BMD [a]			
Lumbar spine (g/cm^2)	0.86 ± 0.12	0.82 ± 0.17	0.15
Femoral neck (g/cm^2)	0.659 ± 0.07	0.64 ± 0.08	0.63
Total hip (g/cm^2)	0.75 ± 0.12	0.81 ± 0.08	0.06
AST (U/L) [a]	23.23 ± 5.96	23 ± 3.17	0.86
ALT (U/L) [a]	23.11 ± 5.75	22.38 ± 7.46	0.69
Cr (mg/dL) [a]	0.71 ± 0.15	0.67 ± 0.14	0.24
TSH (mIU/L) [a]	1.72 ± 0.921	1.44 ± 0.81	0.24
PTH (pg/mL) [a]	44.5 ± 11.6	51.9 ± 16.57	0.07
25(OH)D (ng/mL) [a]	34.48 ± 8.08	35.18 ± 10.83	0.79

Notes: [a] Data expressed as mean ± standard deviation (SD), [b] Data expressed as a percentage. A statistically significant difference in the groups was p-value < 0.05. BMD, bone mineral density; AST, aspartate transaminase; ALT, alanine transaminase; Cr, creatinine; TSH, Thyroid-stimulating hormone; PTH, Parathyroid hormone; 25(OH)D, 25-hydroxyvitamin D.

Table 2. Comparison of the bone turnover marker between the vitamin E group and placebo group at baseline and 12 weeks.

Bone Turnover Marker	Vitamin E (n = 26)	Placebo (n = 26)	p-Value
CTX (ng/mL)			
Baseline [a]	0.39 (0.10, 0.66)	0.28 (0.10, 1.02)	0.17
12 weeks [a]	0.40 (0.02, 0.66)	0.45 (0.19, 1.16)	0.78
Mean difference [b]	−0.003 ± 0.09	0.121 ± 0.15	<0.001 *
P1NP (ng/mL)			
Baseline [a]	50.80 (6.96, 78.90)	52.02 (20.52, 127.9)	0.78
12 weeks [a]	50.66 (20.43, 78.90)	49.85 (21.17, 139.4)	0.78
Mean difference [a]	−2.75 (−28.82, 48.41)	4.87 (−27.09, 19.93)	0.10

Notes: [a] Data expressed as median(range), [b] Data expressed as mean ± standard deviation (SD). CTX, C-terminal telopeptide of type I collagen; P1NP, N-terminal propeptide of type I procollagen. * p-value < 0.05 assigned as statistically significant.

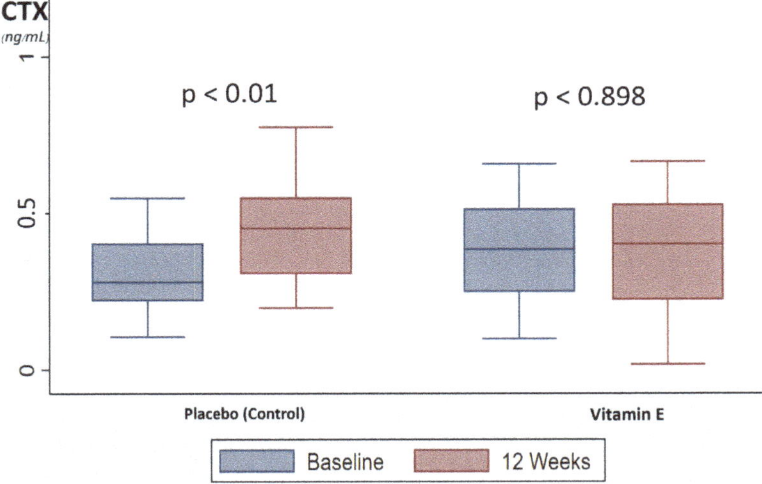

Figure 2. The box-plot represents the changed of bone resorption marker. CTX, C-terminal telopeptide of type I collagen.

3.3. Adverse Events

One participant in the vitamin E group had postmenopausal bleeding after 10 weeks of supplementation and was spontaneously relieved. At the 12-week follow-up visit, this adverse event was self-reported by the participants. She had not had any abnormal bleeding in other sites. A pelvic examination was performed, and the results were unremarkable. Transrectal sonography showed 5 mm of endometrial thickness. Her Pap smear was negative for malignancy. Her laboratory investigation showed no abnormalities, including platelet count, hemoglobin level, coagulogram, and renal and kidney function. Fractional curettage was done, and atrophic endometrium was reported. Neither AST nor ALT level was affected after 12 weeks of study in any participants.

4. Discussion

A vitamin E, or mixed tocopherol, supplement for twelve weeks can slow down bone resorption in osteopenic postmenopausal women. This is in line with the animal model study, which showed that vitamin E could suppress bone resorption activity and decrease the rate of bone loss [34,35]. Norazlina et al. studied the effect of vitamin E supplement or alpha-tocopherol on the bone of ovariectomized female rats. They revealed that vitamin E and alpha-tocopherol could suppress bone resorption markers compared to placebo [27].

In addition, Johnson et al. have found that vitamin E supplements can decrease the osteoclast number of femoral bone marrow in ovariectomized rats compared to placebo [32]. This supports the hypothesis that vitamin E (both the tocopherols and tocotrienols) can slow down osteoclast activity exacerbated by estrogen deficiency. This hypothesis is also applicable to humans. Michaëlsson et al. [39] have reported a cohort study of the relationship between intake and serum concentrations of alpha-tocopherol and fractures in 61,433 older Swedish women. The results showed that low intake and low serum concentration of α-tocopherol was associated with an increased fracture rate. Moreover, Holvik et al. investigated the association between serum α-tocopherol concentrations and risk of hip fracture during up to 11 years of follow-up based on the Norwegian Epidemiologic Osteoporosis Studies (NOREPOS) cohort study, which included 21,774 men and women aged between 65 and 79 years. They reported an association between low serum concentrations of vitamin E and increased risk of hip fracture [29].

As mentioned earlier, there is an association between low serum vitamin E or toco-pherol concentration and the risk of osteoporotic fracture in the elderly. To the best of our knowledge, this is the first randomized controlled trial to evaluate mixed-tocopherol sup-plement effects on the bone turnover marker in postmenopausal women with osteopenia. In addition, we have found that vitamin E or mixed-tocopherol has some osteoprotective ef-fect against the physiologic bone loss across menopausal women. The study confirmed that bone resorption maker, or serum CTX, was maintained in the intervention group, whereas in the placebo group, this marker was increased. Moreover, there is a significant difference between the intervention and the control group in terms of the mean difference. The serum CTX has a trend of decrease in the treatment group: decreased bone resorption marker. In contrast, a significant increase in serum CTX was found in the placebo group. This evidence the anti-resorptive effect of vitamin E on the bone remodeling process. However, an increase in serum CTX in the placebo group at 12 weeks was within the Asian variation in CTX changes, which have reported around 37–42% [40]. Rathnayake et al. reported a systemic review of the bone marker turnover in Asian pre-and postmenopausal women, including 23 studies from China, Japan, India, Korea, Pakistan, and Thailand. The results of these studies showed that the mean serum CTX varied from 0.25 to 0.433 ng/mL (42%) and, within China, 0.25–0.395 ng/mL (37%). While Western study, Cavalier et al. conducted the cohort study of European Biological Variation Study; EuBIVAS, which included six European laboratories from Italy, Norway, Spain, the Netherlands, and Turkey. The results showed the variation of serum CTX among Caucasian women, aged more than 50 years varied from 13.4–45.0 ng/mL (21%) [41]. For this evidence, the difference in genetic, ethnics, lifestyle, and dietary culture of dairy products, that affect bone remodeling could be the reasons for the variation of bone markers between Asian and Caucasians.

Tocopherol is a potent antioxidant that is extensively used for wellness supplemen-tation. It can decrease ROS and pro-inflammatory processes. Estrogen deficiency in postmenopausal women increases oxidative stress. The raising of ROS induces osteoclasto-genesis and stimulates osteoclast activity that causes bone loss [6]. Tocopherol supplement can lower ROS, thus suppressing osteoclastogenesis, osteoclast activity, and osteocyte apop-tosis [9,23]. Tocopherol supplementation can also decrease pro-inflammatory cytokines, especially IL-1, IL-6, and TNF-α, which are essential factors of bone resorption in vivo and in vitro. These cytokines stimulate the formation and activity of osteoclasts, leading to excessive bone resorption [3,28,42]. Tocopherol may decrease bone resorption via these two processes.

For the BTMs measurement, we used CTX and PINP, the reference BTMs, recom-mended by the International osteoporosis foundation, and the International federation of clinical chemistry and laboratory medicine's bone marker standard working group as the standard bone formation and bone resorption marker in clinical use. In addition, our measurement method was the automated electrochemiluminescence, which has the least analytical variability compared to other methods, such as enzyme-linked immunosor-bent assay and radioimmunoassay. However, the treatment with an anti-resorptive agent resulted in an early decrease of bone resorption marker at around 3 months, followed by a reduction of bone formation marker in approximately 6 months [37,43]. Results in line with these have been achieved using several anti-resorptive agents. Those results showed a reduction in serum CTX levels at 3 months of administration include alendronate, ibandronate, risedronate, zoledronic acid, raloxifene, hormone replacement therapy, and denosumab. Moreover, a reduction in bone turnover marker has been demonstrated as an independent predictor of fracture risk reduction [37]. Hochberg et al. reported a meta-analysis of 16 randomized trials of anti-resorptive therapy and the extent of change in BMD and decreased bone turnover markers in postmenopausal women with osteoporosis. They reported an association between reduced risk of non-vertebral fractures, increased BMD, reduced bone resorption marker, or serum CTX. The meta-analysis revealed that the estimated 70% reduction in a bone resorption marker corresponds to a 40% reduction in non-vertebral fracture risk [44]. A daily 400 IU tocopherol supplement seems to have some benefits in altering bone resorption markers compared to placebo, which could benefit fracture risk reduction.

There was no transaminitis in all participants after 12 weeks of supplementation. However, one participant in the vitamin E group had vaginal bleeding at 10 weeks of intervention. The cause of the bleeding was atrophic endometrium, according to pathologic results. In vivo study found that vitamin E has an antiplatelet and anticoagulating function [45–47]. Previous studies found an increased risk of bleeding from vitamin E supplements in patients who take anticoagulants and have vitamin K deficiency. Hence, vitamin E supplementation should be avoided in high-risk groups, given that no such complications were found in a healthy population, even in high dose vitamin E supplement at a dose 2000 IU per day [48–50].

Managing osteopenia in postmenopausal women to prevent osteoporosis and fracture includes: calcium and vitamin D supplements, appropriate protein intake, weight-bearing exercise, and avoiding risk factors, such as smoking alcohol, caffeine, and falls [18,51]. This study shows that adding vitamin E supplements combined with calcium plus vitamin D can decrease bone resorption, benefiting postmenopausal osteopenic women.

The strength of this study is its design being a randomized, double-blind, placebo-controlled clinical trial that uses intention-to-treat analysis. Most participants were of good nutritional status, as indicated by the average BMI of 22 kg/m^2 and had high compliance with medication intake and completed follow-up in both groups. Adverse effects were evaluated, including signs, symptoms, and liver function tests.

The follow-up duration was not long enough to detect a significant change of bone formation marker (P1NP). In addition, the baseline vitamin E level of the participants, daily dietary vitamin E intake, and lifestyle, such as exercise, were not evaluated.

We suggest a longer duration of follow-up and different doses of mixed-tocopherol for supplementation in a future study. Bone density or fracture events would be informative to evaluate the effect of vitamin E on osteoporotic fracture prevention.

5. Conclusions

In conclusion, vitamin E (mixed-tocopherol) supplementation in postmenopausal osteopenic women slows down the increase of bone resorption marker (CTX) that may represent bone loss prevention through anti-resorptive activity. No significant adverse effect was found after 12 weeks of supplementation.

Author Contributions: S.A.-O.V., K.N. and O.V. conceived the research question and hypothesis. K.N., C.T. and O.V. conducted the data collection process. S.A.-O.V., K.N. and O.V. checked and validated of data. S.A.-O.V., K.N. and O.V. conducted the statistical analyses and provided insights of its clinical significance. S.A.-O.V., K.N., O.V. drafting the manuscript. S.A.-O.V., K.N., A.S., O.V. and especially S.A.-O.V. and O.V. contributed critical revision of the manuscript for important intellectual content. S.A.-O.V. and O.V. final approval of the version to be published. All authors have read and agreed to the published version of the manuscript.

Funding: This research received from the Faculty of Medicine, Ramathibodi Hospital, no external funding.

Institutional Review Board Statement: The study was conducted according to the guidelines of the Declaration of Helsinki, and approved by the Ethical Clearance Committee on Human Rights Related to Research Involving Human Subjects Faculty of Medicine Ramathibodi Hospital, Mahidol University (MURA 2018/02), and study protocol also submitted to the Thai Clinical Trials Registry; TCTR (www.clinicaltrials.in.th) Clinical trial registration no. TCTR20180419001/18-04-2018.

Informed Consent Statement: Informed consent was obtained from all subjects involved in the study.

Data Availability Statement: Data available on request due to restrictions on privacy or ethics. The data presented in this study are available on request from the corresponding author. The data are not publicly available due to the ethics and right of the faculty of medicine, Ramathibodi Hospital, Mahidol University.

Conflicts of Interest: The authors declare no conflict of interest.

Abbreviations

25(OH)D = 25-hydroxyvitamin D; AFF = atypical femoral fracture; ALT = alanine aminotransferase; AST = aspartate aminotransferase; B-ALP = bone-specific alkaline phosphatase; BMD = bone mineral density; BMU = bone multicellular unit; BTMs = bone turnover markers; CBC = complete blood count; CTX = C-terminal telopeptide; CTX-1 = carboxy-terminal cross-linked telopeptide of type 1 collagen; DPD = deoxypyridinoline; DXA = dual-energy X-ray absorptiometry; FRAX$^{®}$ = fracture risk assessment tool; HYP = hydroxyproline; IL-1 = Interleukin-1; IL-6 = Interleukin-6; NTX-1 = amino-terminal cross-linked telopeptide of type 1 collagen; O = osteocalcin; ONJ = osteonecrosis of jaw; OPG = osteoprotegerin; P1CP = procollagen type 1 C-terminal propeptide; P1NP = procollagen type 1 N-terminal propeptide; PGE2 = prostaglandin E2; PTH = parathyroid hormone; RANK = the receptor activator of nuclear factor kappa-B receptor activator of nuclear factor kappa-B ligand; ROS = reactive oxygen species; T3 = tocotrienol; TF = tocopherol; TNF-α = tumor necrotic factor α; TRAP 5b = tartrate-resistant acid phosphatase 5b; TSH = thyroid-stimulating hormone.

References

1. Baron, R.; Hesse, E. Update on bone anabolics in osteoporosis treatment: Rationale, current status, and perspectives. *J. Clin. Endocrinol. Metab.* **2012**, *97*, 311–325. [CrossRef] [PubMed]
2. Siddiqui, J.A.; Partridge, N.C. Physiological Bone Remodeling: Systemic Regulation and Growth Factor Involvement. *Physiology* **2016**, *31*, 233–245. [CrossRef]
3. Kanzaki, H.; Shinohara, F.; Kanako, I.; Yamaguchi, Y.; Fukaya, S.; Miyamoto, Y.; Wada, S.; Nakamura, Y. Molecular regulatory mechanisms of osteoclastogenesis through cytoprotective enzymes. *Redox Biol.* **2016**, *8*, 186–191. [CrossRef]
4. Kenkre, J.S.; Bassett, J.H.D. The bone remodelling cycle. *Ann. Clin. Biochem.* **2018**, *55*, 308–327. [CrossRef] [PubMed]
5. Bai, X.C.; Lu, D.; Bai, J.; Zheng, H.; Ke, Z.Y.; Li, X.M.; Luo, S.Q. Oxidative stress inhibits osteoblastic differentiation of bone cells by ERK and NF-κB. *Biochem. Biophys. Res. Commun.* **2004**, *314*, 197–207. [CrossRef]
6. Lean, J.M.; Jagger, C.J.; Kirstein, B.; Fuller, K.; Chambers, T.J. Hydrogen Peroxide Is Essential for Estrogen-Deficiency Bone Loss and Osteoclast Formation. *Endocrinology* **2005**, *146*, 728–735. [CrossRef]
7. Allshouse, A.; Pavlovic, J.; Santoro, N. Menstrual Cycle Hormone Changes Associated with Reproductive Aging and How They May Relate to Symptoms. *Obstet. Gynecol. Clin. N. Am.* **2018**, *45*, 613–628. [CrossRef] [PubMed]
8. Tella, S.H.; Gallagher, J.C. Prevention and treatment of postmenopausal osteoporosis. *J. Steroid Biochem. Mol. Biol.* **2014**, *142*, 155–170. [CrossRef]
9. Baek, K.H.; Oh, K.W.; Lee, W.Y.; Lee, S.S.; Kim, M.K.; Kwon, H.S.; Rhee, E.J.; Han, J.H.; Song, K.H.; Cha, B.Y.; et al. Association of Oxidative Stress with Postmenopausal Osteoporosis and the Effects of Hydrogen Peroxide on Osteoclast Formation in Human Bone Marrow Cell Cultures. *Calcif. Tissue Int.* **2010**, *87*, 226–235. [CrossRef]
10. Harrison, B.J.; Hutchinson, C.E.; Adams, J.; Bruce, I.N.; Herrick, A. Assessing periarticular bone mineral density in patients with early psoriatic arthritis or rheumatoid arthritis. *Ann. Rheum. Dis.* **2002**, *61*, 1007–1011. [CrossRef]
11. Mateen, S.; Moin, S.; Khan, A.Q.; Zafar, A.; Fatima, N. Increased Reactive Oxygen Species Formation and Oxidative Stress in Rheumatoid Arthritis. *PLoS ONE* **2016**, *11*, e0152925. [CrossRef] [PubMed]
12. Report of a WHO Study Group. Assessment of fracture risk and its application to screening for postmenopausal osteoporosis. *World Health Organ. Tech. Rep. Ser.* **1994**, *843*, 1–129.

13. Saengsuda, S. Prevalence of Osteoporosis and Osteopenia in Thai Female Patients at Rajavithi Hospital. *J. Health Sci.* **2017**, *22*, 242–250.

14. Wright, N.C.; Looker, A.C.; Saag, K.G.; Curtis, J.R.; Delzell, E.S.; Randall, S.; Dawson-Hughes, B. The Recent Prevalence of Osteoporosis and Low Bone Mass in the United States Based on Bone Mineral Density at the Femoral Neck or Lumbar Spine. *J. Bone Miner. Res.* **2014**, *29*, 2520–2526. [CrossRef]

15. Kaushal, N.; Vohora, D.; Jalali, R.K.; Jha, S. Prevalence of osteoporosis and osteopenia in an apparently healthy Indian population— A cross-sectional retrospective study. *Osteoporos. Sarcopenia* **2018**, *4*, 53–60. [CrossRef] [PubMed]

16. Pasco, J.A.; Seeman, E.; Henry, M.J.; Merriman, E.N.; Nicholson, G.C.; Kotowicz, M.A. The population burden of fractures originates in women with osteopenia, not osteoporosis. *Osteoporos. Int.* **2006**, *17*, 1404–1409. [CrossRef] [PubMed]

17. Wattanachanya, L.; Pongchaiyakul, C. Prevalence and risk factors of morphometric vertebral fracture in apparently healthy osteopenic postmenopausal Thai women. *Menopause* **2020**, *28*, 12–17. [CrossRef]

18. Camacho, P.M.; Petak, S.M.; Binkley, N.; Diab, D.L.; Eldeiry, L.S.; Farooki, A.; Harris, S.T.; Hurley, D.L.; Kelly, J.; Lewiecki, E.M.; et al. American association of clinical endocrinologists/american college of endocrinology clinical practice guidelines for the diagnosis and treatment of postmenopausal osteoporosis-2020 update. *Endocr. Pract.* **2020**, *26*, 1–46. [CrossRef] [PubMed]

19. Reid, I.R.; Cundy, T. Osteonecrosis of the jaw. *Skeletal Radiol.* **2009**, *38*, 5–9. [CrossRef] [PubMed]

20. Nicolatou-Galitis, O.; Schiødt, M.; Mendes, R.A.; Ripamonti, C.; Hope, S.; Drudge-Coates, L.; Niepel, D.; Van den Wyngaert, T. Medication-related osteonecrosis of the jaw: Definition and best practice for prevention, diagnosis, and treatment. *Oral Surg. Oral Med. Oral Pathol. Oral Radiol.* **2019**, *127*, 117–135. [CrossRef]

21. Khow, K.S.F.; Shibu, P.; Yu, S.C.Y.; Chehade, M.J.; Visvanathan, R. Epidemiology and postoperative outcomes of atypical femoral fractures in older adults: A systematic review. *J. Nutr. Health Aging* **2017**, *21*, 83–91. [CrossRef] [PubMed]

22. Silverman, S.; Kupperman, E.; Bukata, S. Bisphosphonate-related atypical femoral fracture: Managing a rare but serious complication. *Clevel. Clin. J. Med.* **2018**, *85*, 885–893. [CrossRef] [PubMed]

23. Wong, S.K.; Mohamad, N.-V.; Ibrahim, N.; Chin, K.-Y.; Shuid, A.N.; Ima-Nirwana, S. The Molecular Mechanism of Vitamin E as a Bone-Protecting Agent: A Review on Current Evidence. *Int. J. Mol. Sci.* **2019**, *20*, 1453. [CrossRef]

24. Domazetovic, V.; Marcucci, G.; Iantomasi, T.; Brandi, L.; Vincenzin, M. Oxidative stress in bone remodeling: Role of antioxidants. *Clin. Cases Miner. Bone Metab.* **2017**, *14*, 209–216. [CrossRef] [PubMed]

25. Chin, K.Y.; Ima-Nirwana, S. The effects of alpha-tocopherol on bone: A double-edged sword? *Nutrients* **2014**, *6*, 1424–1441. [CrossRef]

26. Naina Mohamed, I.; Borhanuddin, B.; Shuid, A.N.; Mohd Fozi, N.F. Vitamin E and Bone Structural Changes: An Evidence-Based Review. *Evid.-Based Complement. Altern. Med.* **2012**, *2012*, 250584. [CrossRef]

27. Norazlina, M.; Ima-Nirwana, S.; Gapor, M.T.; Khalid, B.A.K. Palm vitamin E is comparable to α-tocopherol in maintaining bone mineral density in ovariectomised female rats. *Exp. Clin. Endocrinol. Diabetes* **2000**, *108*, 305–310. [CrossRef]

28. Nazrun, A.S.; Norazlina, M.; Norliza, M.; Nirwana, S.I. The Anti-Inflammatory Role of Vitamin E in Prevention of Osteoporosis. *Adv. Pharmacol. Sci.* **2012**, *2012*, 142702. [CrossRef]

29. Holvik, K.; Gjesdal, C.G.; Tell, G.S.; Grimnes, G.; Schei, B.; Apalset, E.M.; Samuelsen, S.O.; Blomhoff, R.; Michaëlsson, K.; Meyer, H.E. Low serum concentrations of alpha-tocopherol are associated with increased risk of hip fracture. A NOREPOS study. *Osteoporos. Int.* **2014**, *25*, 2545–2554. [CrossRef]

30. Mata-Granados, J.M.; Cuenca-Acebedo, R.; de Castro, M.D.L.; Gómez, J.M.Q. Lower vitamin E serum levels are associated with osteoporosis in early postmenopausal women: A cross-sectional study. *J. Bone Miner. Metab.* **2013**, *31*, 455–460. [CrossRef]

31. Iwaniec, U.T.; Turner, R.T.; Smith, B.; Stoecker, B.J.; Rust, A.; Zhang, B.; Vasu, V.T.; Gohil, K.; Cross, C.E.; Traber, M.G. Evaluation of long-term vitamin E insufficiency or excess on bone mass, density, and microarchitecture in rodents. *Free. Radic. Biol. Med.* **2013**, *65*, 1209–1214. [CrossRef]

32. Johnson, S.A.; Feresin, R.G.; Soung, D.Y.; Elam, M.L.; Arjmandi, B.H. Vitamin E suppresses ex vivo osteoclastogenesis in ovariectomized rats. *Food Funct.* **2016**, *7*, 1628–1633. [CrossRef]

33. Muhammad, N.; Luke, D.A.; Shuid, A.N.; Mohamed, N.; Soelaiman, I.-N. Two Different Isomers of Vitamin E Prevent Bone Loss in Postmenopausal Osteoporosis Rat Model. *Evid.-Based Complement. Altern. Med.* **2012**, *2012*, 161527. [CrossRef] [PubMed]

34. Reagan-Shaw, S.; Nihal, M.; Ahmad, N. Dose translation from animal to human studies revisited. *FASEB J.* **2008**, *22*, 659–661. [CrossRef]

35. Feresin, R.G.; Johnson, S.A.; Elam, M.L.; Kim, J.-S.; Khalil, D.A.; Lucas, E.A.; Smith, B.J.; Payton, M.E.; Akhter, M.P.; Arjmandi, B.H. Effects of Vitamin E on Bone Biomechanical and Histomorphometric Parameters in Ovariectomized Rats. *J. Osteoporos.* **2013**, *2013*, 825985. [CrossRef]

36. Civitelli, R.; Armamento-Villareal, R.; Napoli, N. Bone turnover markers: Understanding their value in clinical trials and clinical practice. *Osteoporos. Int.* **2009**, *20*, 843–851. [CrossRef] [PubMed]

37. Vasikaran, S.; Cooper, C.; Eastell, R.; Griesmacher, A.; Morris, H.A.; Trenti, T.; Kanis, J.A. International Osteoporosis Foundation and International Federation of Clinical Chemistry and Laboratory Medicine Position on bone marker standards in osteoporosis. *Clin. Chem. Lab. Med.* **2011**, *49*, 1271–1274. [CrossRef]

38. Eastell, R.; Szulc, P. Use of bone turnover markers in postmenopausal osteoporosis. *Lancet Diabetes Endocrinol.* **2017**, *5*, 908–923. [CrossRef]

39. Michaëlsson, K.; Wolk, A.; Byberg, L.; Ärnlöv, J.; Melhus, H. Intake and serum concentrations of α-tocopherol in relation to fractures in elderly women and men: 2 cohort studies. *Am. J. Clin. Nutr.* **2014**, *99*, 107–114. [CrossRef]
40. Rathnayake, H.; Lekamwasam, S.; Wickramatilake, C.; Lenora, J. Variation of urinary and serum bone turnover marker reference values among pre and postmenopausal women in Asia: A systematic review. *Arch. Osteoporos.* **2020**, *15*, 57. [CrossRef]
41. Cavalier, E.; Lukas, P.; Bottani, M.; Aarsand, A.K.; Ceriotti, F.; Coşkun, A.; Díaz-Garzón, J.; Fernàndez-Calle, P.; Guerra, E.; Locatelli, M.; et al. European Biological Variation Study (EuBIVAS): Within- and between-subject biological variation estimates of β-isomerized C-terminal telopeptide of type I collagen (β-CTX), N-terminal propeptide of type I collagen (PINP), osteocalcin, intact fibroblast growth factor 23 and uncarboxylated-unphosphorylated matrix-Gla protein-a cooperation between the EFLM Working Group on Biological Variation and the International Osteoporosis Foundation-International Federation of Clinical Chemistry Committee on Bone Metabolism. *Osteoporos. Int.* **2020**, *31*, 1461–1470.
42. Belisle, S.E.; Leka, L.S.; Dallal, G.E.; Jacques, P.F.; Ordovas, J.; Meydani, S.N.; Delgado-Lista, J. Cytokine response to vitamin E supplementation is dependent on pre-supplementation cytokine levels. *BioFactors* **2008**, *33*, 191–200. [CrossRef]
43. Morris, H.A.; Eastell, R.; Jørgensen, N.R.; Cavalier, E.; Vasikaran, S.; Chubb, S.A.P.; Kanis, J.A.; Cooper, C.; Makris, K. Clinical usefulness of bone turnover marker concentrations in osteoporosis. *Clin. Chim. Acta* **2017**, *467*, 34–41. [CrossRef]
44. Hochberg, M.C.; Greenspan, S.; Wasnich, R.D.; Miller, P.; Thompson, D.E.; Ross, P.D. Changes in bone density and turnover explain the reductions in incidence of nonvertebral fractures that occur during treatment with antiresorptive agents. *J. Clin. Endocrinol. Metab.* **2002**, *87*, 1586–1592. [CrossRef] [PubMed]
45. Glynn, R.J.; Ridker, P.M.; Goldhaber, S.Z.; Zee, R.Y.; Buring, J.E. Effects of random allocation to vitamin E supplementation on the occurrence of venous thromboembolism: Report from the Women's Health Study. *Circulation* **2007**, *116*, 1497–1503. [CrossRef] [PubMed]
46. Steiner, M. Influence of vitamin E on platelet function in humans. *J. Am. Coll. Nutr.* **1991**, *10*, 466–473. [CrossRef] [PubMed]
47. Vignini, A.; Nanetti, L.; Moroni, C.; Testa, R.; Sirolla, C.; Marra, M.; Manfrini, S.; Fumelli, D.; Marcheselli, F.; Mazzanti, L.; et al. A study on the action of vitamin E supplementation on plasminogen activator inhibitor type 1 and platelet nitric oxide production in type 2 diabetic patients. *Nutr. Metab. Cardiovasc. Dis.* **2008**, *18*, 15–22. [CrossRef]
48. Violi, F.; Pignatelli, P.; Basili, S.; Grum, F.; Hufendiek, K.; Franz, S.; Bogdahn, U.; Gamulescu, M.; Rümmele, P.; Schlachetzki, F. Nutrition, Supplements, and Vitamins in Platelet Function and Bleeding. *Circulation* **2010**, *121*, 1033–1044. [CrossRef]
49. Vatassery, G.T.; Bauer, T.; Dysken, M. High doses of vitamin E in the treatment of disorders of the central nervous system in the aged. *Am. J. Clin. Nutr.* **1999**, *70*, 793–801. [CrossRef]
50. Abner, E.L.; Schmitt, F.A.; Mendiondo, M.S.; Marcum, J.; Kryscio, R.J. Vitamin E and All-Cause Mortality: A Meta-Analysis. *Curr. Aging Sci.* **2011**, *4*, 158–170. [CrossRef]
51. Songpatanasilp, T.; Sritara, C.; Kittisomprayoonkul, W.; Chaiumnuay, S.; Nimitphong, H.; Charatcharoenwitthaya, N.; Pongchaiyakul, C.; Namwongphrom, S.; Kitumnuaypong, T.; Srikam, W.; et al. Thai Osteoporosis Foundation (TOPF) position statements on management of osteoporosis. *Osteoporos. Sarcopenia* **2016**, *2*, 191–207. [CrossRef] [PubMed]

Article

Effect of Vitamin D-Enriched Gouda-Type Cheese Consumption on Biochemical Markers of Bone Metabolism in Postmenopausal Women in Greece

George Moschonis [1,*], Ellen GHM van den Heuvel [2], Christina Mavrogianni [3] and Yannis Manios [3]

[1] Department of Dietetics, Nutrition and Sport, School of Allied Health, Human Services and Sport, La Trobe University, Melbourne 3086, Australia
[2] FrieslandCampina, Stationsplein 4, Post Box 1551, 3800 BN Amersfoort, The Netherlands; eghm.vandenheuvel@outlook.com
[3] Department of Nutrition and Dietetics, School of Health Science and Education, Harokopio University, 17671 Athens, Greece; cmavrog@hua.gr (C.M.); manios@hua.gr (Y.M.)
* Correspondence: g.moschonis@latrobe.edu.au; Tel.: +61-3-9479-3482

Citation: Moschonis, G.; van den Heuvel, E.G.; Mavrogianni, C.; Manios, Y. Effect of Vitamin D-Enriched Gouda-Type Cheese Consumption on Biochemical Markers of Bone Metabolism in Postmenopausal Women in Greece. *Nutrients* **2021**, *13*, 2985. https://doi.org/10.3390/nu13092985

Academic Editor: LaVerne L. Brown

Received: 19 July 2021
Accepted: 26 August 2021
Published: 27 August 2021

Abstract: Considering the role of bone metabolism in understanding the pathogenesis of osteoporosis, the aim of the present study was to examine the effects of vitamin D-enriched cheese on the serum concentrations of the parathyroid hormone (PTH) and certain bone remodeling biomarkers in postmenopausal women in Greece. In a randomised, controlled dietary intervention, 79 postmenopausal women (55–75 years old) were randomly allocated either to a control (CG: n = 39) or an intervention group (IG: n = 40), consuming 60 g of either non-enriched or vitamin D3-enriched Gouda-type cheese (5.7 µg of vitamin D3), respectively, daily and for eight weeks during the winter. The serum concentrations of 25-hydroxy vitamin D (25(OH)D), PTH, bone formation (i.e., osteocalcin, P1NP) and bone resorption (i.e., TRAP-5b) biomarkers were measured. Consumption of the vitamin D-enriched cheese led to higher serum 25(OH)D concentrations of 23.4 ± 6.39 (p = 0.022) and 13.4 ± 1.35 (p < 0.001) nmol/L in vitamin D-insufficient women being at menopause for less and more than 5 years, respectively. In vitamin D-insufficient women that were less than 5 years at menopause, consumption of vitamin D-enriched cheese was also associated with lower serum PTH (Beta −0.63 ± 1.11; p < 0.001) and TRAP-5b (Beta −0.65 ± 0.23; p = 0.004) levels at follow-up, compared with the CG. The present study showed that daily intake of 5.7 µg of vitamin D through enriched cheese increased serum 25(OH)D concentrations, prevented PTH increase and reduced bone resorption in vitamin D-insufficient early postmenopausal women, thus reflecting a potential food-based solution for reducing the risk of bone loss occurring after menopause.

Keywords: postmenopausal women; vitamin D; bone remodeling; cheese; enriched dairy

1. Introduction

In 2010, the US Institute of Medicine (IOM) released the latest and most updated Dietary Reference Intake (DRI) for vitamin D, which was mainly based on evidence indicating the beneficial effects of vitamin D on musculoskeletal health outcomes, such as fractures, osteomalacia, rickets, muscle weakness, falls, etc. [1]. The DRIs of vitamin D have been also established based on the effects of dietary vitamin D intake on serum 25-hydroxy vitamin D (25(OH)D) concentrations and the intention of keeping them at levels equal or above 50 nmol/L, which is the diagnostic threshold for vitamin D sufficiency [2]. This is because serum concentrations of 25(OH)D of at least 50 nmol/L are the basic requirement for the normalisation of parathyroid hormone (PTH) levels in blood [3]. As both vitamin D and PTH are involved in controlling calcium and bone metabolism, the clinical outcomes of vitamin D insufficiency (i.e., 25(OH)D < 50 nmol/L) and the consequent secondary hyperparathyroidism in the skeleton is an accelerated rate of bone remodeling, which if sustained gradually leads to bone loss [3]. Bone remodeling describes a dynamic process that is reflected in the interrelated function of bone-forming osteoblasts and bone-resorbing osteoclasts [4,5]. During bone resorption, osteoclastic activity leads to the release of breakdown products of type-I collagen, a process that is usually mediated by certain enzymes, such as the osteoclast-specific 5b isoform of tartrate-resistant acid phosphatase (TRAP-5b) [6]. In healthy bone, the resorption cavity created by osteoclasts is filled with new osteoid material secreted by active osteoblasts. As part of this process, one of the bone formation molecules secreted by active osteoblasts is osteocalcin (OC), which binds to the mineralised bone matrix [7]. Another typical bone formation biomarker is type I procollagen-N-propeptide (P1NP), which is a byproduct of collagen synthesis. More specifically, following the synthesis of new type I collagen within the osteoblast, P1NP is cleaved from type I procollagen by proteases outside the osteoblast [6,8].

Women lose on average about 2% to 2.5% of their bone during the transmenopause period, which is one year before and two years after the last menstrual period [9]. Nevertheless, the rapid decline in bone mineral density (BMD) occurs mainly after menopause, and as such, the cumulative 10-year lumbar spine BMD loss reported for Caucasian women is 10.6% [10], with most of it lost within the first 3 years after menopause, thus highlighting a time-period during which the risk of osteoporotic fractures is elevated [11]. Besides menopause in women, some of the main risk factors for the progression and development of bone loss are also related to lifestyle and particularly to inadequate dietary intake of certain essential micronutrients [12]. The combined intake of both vitamin D and calcium results in a significant reduction in non-vertebral fractures, especially in those individuals with insufficient vitamin D status [13,14]. Even though there is an abundance of clinical trials indicating the beneficial effects of calcium and/or vitamin D supplementation on the prevention of bone loss and on the levels of bone remodeling biomarkers [15–19], there are extremely few dietary intervention studies examining the effect of increased intake of these micronutrients through fortified foods [20]. Previously, we showed that consumption of 60 g of reduced-fat Gouda-type cheese enriched with vitamin D3 was effective in decreasing the prevalence of vitamin D insufficiency during winter months in a population of postmenopausal women in Greece [21].

Considering the emerging role of bone metabolism in understanding the pathogenesis of osteoporosis, the primary aim of the present study was to examine the effects of vitamin D-enriched cheese on the serum concentrations of PTH and certain bone remodelling biomarkers in early or late postmenopausal women in Greece, with sufficient or insufficient vitamin D status at the start of the intervention.

2. Materials and Methods

2.1. Study Design and Sampling Procedures

The current study was a single-blinded (i.e., blinded only to study participants), randomised controlled dietary intervention study, testing the effect of vitamin D_3-enriched, reduced-fat Gouda-type cheese on the serum concentrations of certain calciotropic hormones (i.e., 25(OH)D, PTH), bone formation (i.e., OC, P1NP) and bone resorption markers (i.e., TRAP-5b) in postmenopausal women.

The sampling procedures followed in the present study, which was originally designed to assess vitamin D status changes as a primary outcome, have been described in detail elsewhere [21]. According to sample size estimation, a sample of 37 women per treatment was adequate to provide statistical power greater than 80% (alpha = 0.05, two-tailed) for detecting an increase in serum 25(OH)D levels of ~6 nmol/L from baseline to follow-up at probability of type I error <0.05. To account for potential dropouts, the number of women was increased to 40 per group.

2.2. Screening and Ethics Approval

The study was initiated with two screening phases in October 2014, when volunteers were invited to participate via informational brochures and posters which were distributed in public buildings and community centres in two municipalities within the wider district of Athens, namely Kallithea and Tavros-Moschato. Women eligible to participate in the study were those that had no disease/pathology that interacts with vitamin D metabolism; those not requiring or taking any medications, (including hormone replacement therapy) that interact with vitamin D metabolism or vitamin D supplements for medical reasons (e.g., osteoporosis); those not having cow's milk allergy or a history of drug and/or alcohol abuse; those who used to consume cheese daily; those that had not a planned vacation to a sunny holiday destination during the intervention period; and those who were more than 5 years post-menopause. However, the screening process resulted in a small number of women that were less than 5 years since menopause, but as they were close to the 5-year threshold and were satisfying all other inclusion criteria they were also included in the study. Through the first screening phase, 135 women (aged 55–75 y) satisfying the inclusion criteria were identified and were invited to participate in the second screening (December 2014).

The second screening was carried out via scheduled meetings and personal sessions with each one of the 135 eligible women at the Metabolic Unit of the Laboratory of Nutrition and Clinical Dietetics (LNCD) at Harokopio University. As described in detail previously [21], this second screening phase yielded 80 eligible women having, for the most part, homogenous characteristics at baseline. All study participants signed written informed consent forms. The study was approved by the Ethics Committee of Harokopio University (Approval code: 43/23-07-2014) and was conducted in accordance with the code of ethics of the World Medical Association (Declaration of Helsinki) for experiments involving humans. The study protocol registration number was ClinicalTrials.gov: NCT02543671.

2.3. Treatments and Intervention

The 80 eligible women identified after the second screening phase were randomly allocated to an intervention group (IG) and a control group (CG) using a random sampling approach [21]. Four weeks prior to the initiation of the intervention, any eligible participants reporting the use of Vitamin D dietary supplements were instructed to discontinue any intake of them (i.e., one woman in the CG and one woman in the IG). This washout period was included prior to the initiation of the intervention, in order to avoid or minimise any possible effect of supplement use on 25(OH)D concentrations at baseline. Study participants were provided with and were also instructed to consume (as part of their usual diet), 60 g of non-enriched reduced-fat Gouda-type cheese (CG; n = 40) or vitamin D3-enriched, reduced-fat Gouda-type cheese (IG; n = 40) for eight weeks during the winter (i.e., from January to March 2015). Both types of Gouda cheese (i.e., the vitamin D-enriched and the non-enriched type) examined in the present study were made from cow's milk and provided the same amount of several important nutrients, such as protein, phosphorus, calcium, zinc and vitamin B12. However, the 60 g of the vitamin D3-enriched, reduced-fat Gouda-type cheese also provided 5.7 µg of vitamin D per day, which was independently verified using a liquid chromatography–mass spectrometry-based method, as previously described for pork meat [22].

All study participants were instructed to equally replace their habitual consumption of cheese with the corresponding amount of the experimental cheese, so as to compensate for additional calories. In addition, they received detailed guidance by the research group members to report any signs of illness, use of medication and any deviations from the study protocol. Meetings with participants were held biweekly on campus at Harokopio University in Athens, to supply to them the amount of cheese required for the next two weeks. Compliance to the intervention scheme (i.e., adherence to daily consumption of the 60 g of experimental cheese) was also assessed, as described in detail elsewhere [21].

2.4. Measurements

The effectiveness of the intervention was assessed with the assessment of dietary intake and clinical examinations conducted at baseline and follow-up using the same procedures and methodology. The dietary data obtained at the second screening of the study were used as baseline intake data. The follow-up examination took place in mid-March 2015, after eight weeks of intervention.

2.4.1. Dietary Intake

An existing quantitative FFQ validated for calcium intake [23] was updated to also include vitamin D dietary sources (such as fish/seafood, beef, pork, mushrooms, eggs and margarine) and was used to assess habitual dietary calcium and vitamin D intakes. The mean content of the foods listed in the FFQ in calcium and vitamin D was multiplied to their daily frequency of consumption, thus providing the amount of habitual, mean daily calcium and vitamin D intakes.

2.4.2. Anthropometry

Body weight and standing height were measured following the procedures that have been previously described in detail [21]. Body mass index (BMI) (in kg/m^2) was calculated by dividing body weight to standing height squared.

2.4.3. Blood Analyses

Early morning venous blood samples were obtained from each study participant following a 12 h overnight fast at baseline and follow-up examination. Blood was collected in vacutainers without added anticoagulant and was kept at room temperature for $\approx$2 h, where it was allowed to clot. Centrifugation for serum separation was conducted at 3000 rpm for 15 min. Aliquots of 1.5 mL serum were then stored at -80 °C. Serum concentrations of total 25(OH)D (i.e., 25(OH)D$_2$ plus 25(OH)D$_3$) in all serum samples were measured in the Cork Centre for Vitamin D and Nutrition at University College Cork, using a liquid chromatography–tandem mass spectrometry method, described in detail elsewhere [21,24]. Based on the obtained measurements, vitamin D insufficiency and deficiency were defined at serum 25(OH)D concentrations below 50 and 30 nmol/L, respectively [25].

Serum intact PTH and P1NP concentrations were measured with electrochemiluminescence immunoassay (ECLIA) (Roche Diagnostics, Mannheim, Germany) using an ECLIA Elecsys autoanalyser (Cobas e 601). Furthermore, serum total OC concentrations were measured using chemiluminescence immunoassay (CLIA) (Immulite 2000xpi Siemens Healthcare, Erlagen, Germany). In addition, the concentrations of TRAP-5b were measured using an immunocapture enzymatic assay (MetraTRAP-5b EIA Kit, Quidel Corporation, San Diego, CA, USA). The intra-assay and inter-assay coefficients of variation of the biochemical assays performed to measure the concentrations of the aforementioned biochemical indices were: less than 6% for PTH, less than 5% for P1NP; 2.8% and 4.5%, respectively, for osteocalcin; and less than 3% for TRAP-5b.

2.4.4. Statistical Analysis

Descriptive data were reported as mean and standard deviations (SD) for continuous variables or as frequencies (n) and percentages (%) for categorical ones. The Kolmogorov–Smirnov test was used to determine normality of distribution of the examined continuous variables. Between-group differences at baseline were assessed with Student's *t*-test. The chi-squared test was also used to examine between-group differences in the percentages of subjects with serum 25(OH)D levels below 30 and 50 nmol/L at baseline. General estimating equations (GEE) were applied with correction for baseline values in order to examine the effect of treatment with the vitamin D-enriched cheese versus the control group (independent variable) on the serum concentrations of calciotropic hormones and bone remodelling biomarkers at follow-up (dependent variables). Interaction was studied by adding a product term of group (i.e., intervention vs. control group) times vitamin D status (i.e., 25(OH)D < 50 nmol/L vs. 25(OH)D ≥ 50 nmol/L) or years since menopause (i.e., ≤ vs. >5 years). A p-value lower than 0.1 for the interaction term was considered significant. The analysis was stratified by insufficiency or sufficiency in participants' vitamin D status (i.e., 25(OH)D< or ≥50 nmol/L, respectively) and by years since menopause. Subsequently, within each stratum, the other potential effect modifiers were tested. Years since menopause (i.e., ≤ or >5 years) remained significant for PTH and TRAP-5b, and cases of vitamin D sufficiency were also significant for OC. To obtain the regression coefficient for those women that were at menopause for more than 5 years, the coding of the categorial variable was reversed (0 = 1, and 1 = 0). The cut-off of 5 years since menopause was chosen as after this time threshold, the hormone levels in women are more balanced [26]. All models were adjusted for potential confounding factors that changed the regression coefficients of the treatment by 10% or more when added to the linear regression model. The SPSS statistical analysis software for Windows, version 24.0 (SPSS Inc., Chicago, IL, USA) was used to perform all statistical analyses. All statistical tests were two-tailed, and the level of statistical significance was set at $p < 0.05$.

3. Results

3.1. Baseline Characteristics of Study Participants

From the 80 eligible women that were randomly allocated to the two treatments, 79 (i.e., 39 in the control group and 40 in the intervention group) completed the study having full data at both baseline and follow-up examination (Figure 1).

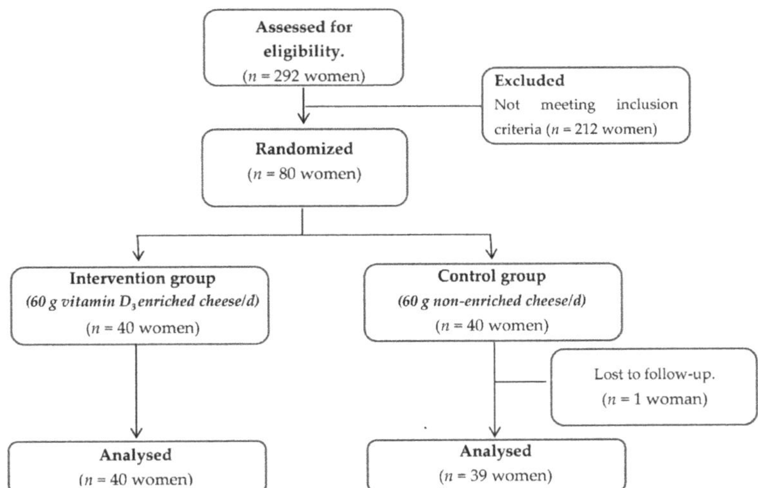

Figure 1. CONSORT flow diagram of study participants.

Table 1 presents the mean sociodemographic characteristics (i.e., age, years since menopause and years of education) of study participants by treatment at baseline, as well as their BMI, serum concentrations of calciotropic hormones (i.e., 25(OH)D and PTH), bone formation (i.e., OC and P1NP) and bone resorption (i.e., TRAP-5b) markers. In addition, Table 1 presents the prevalence of vitamin D deficiency (25(OH)D < 30 nmol/L) and insufficiency (25(OH)D < 50 nmol/L) by treatment arm. No statistically significant differences were observed between the two treatment arms in any of the aforementioned characteristics and blood biomarkers.

Table 1. Baseline descriptive characteristics of study participants completing the trial presented by treatment.

	Intervention Group (n = 40)	Control Group (n = 39)	*p*-Value *
Sociodemographic indices	Mean (SD)		
Age (years)	62.6 (6.0)	63.2 (5.9)	0.670
Years since Menopause	13.9 (5.7)	14.2 (7.3)	0.805
Education (years)	12.3 (3.8)	12.7 (3.5)	0.594
Anthropometric indices			
Body Mass Index (kg/m^2)	28.0 (3.8)	29.0 (2.9)	0.190
Calciotropic hormones			
25(OH)D (nmol/L)	47.3 (15.2)	42.9 (17.7)	0.105
PTH (pg/mL)	41.9 (10.6)	44.2 (13.9)	0.415
Bone formation markers			
Osteocalcin (ng/mL)	24.0 (8.26)	23.6 (7.12)	0.819
P1NP (ng/mL)	50.5 (12.2)	52.5 (14.4)	0.522
Bone resorption markers			
TRAP-5b (U/L)	2.52 (1.03)	2.39 (0.76)	0.533
Vitamin D deficiency	n (%)		
Serum 25(OH)D < 30 nmol/L	5 (12.5)	10 (25.0)	0.137 ‡
Vitamin D insufficiency			
Serum 25(OH)D < 50 nmol/L	27 (67.5)	24 (62.5)	0.750 ‡

* *p*-values were derived from Student's *t*-test, unless stated otherwise. ‡ *p*-values were derived from the chi-squared test.

3.2. Intervention Effect on Serum 25(OH)D Levels in the Total Sample and Vitamin D-Insufficient Women

As described in detail elsewhere [21], cheese consumption resulted in significant increases in dietary calcium intake in both treatment arms, by 191 mg/day (95% CI 103–279) in the IG and 157 mg/day (95% CI 6.7–247) in the CG over the eight weeks, but there was no difference between groups (*p* = 0.6). Similarly, dietary protein intake was equally increased in both treatment arms, since the addition of cheese was the only change introduced in the dietary intake of study participants. Considering that both treatments consumed 60 g of Gouda cheese, protein intake in both the intervention and the control group increased by 18.2 g/day. Mean dietary vitamin D intake increased significantly from baseline to follow-up only in the IG, while it decreased significantly in the CG over the same period, and these changes were also found to be significantly different between groups (5.1 µg (95% CI 4.7–5.5) vs. −0.6 µg (−1.1 to −0.2) *p* < 0.001, respectively). Further to these dietary intake changes, mean serum 25(OH)D concentration increased by 5.1 nmol/L in the IG and decreased by 4.6 nmol/L in the CG during the 8-week intervention period (*p* < 0.001). In addition, 31 out of 40 women in the IG (i.e., 77.5%) responded positively to the implemented intervention by increasing their serum 25(OH)D concentrations by at least 1.08 nmol/L. The increase in serum 25(OH)D concentrations was more pronounced in the vitamin D-insufficient women in the IG, since positive changes were observed for 25 out of 27 (i.e., 92.6%) women consuming the vitamin D-enriched cheese.

3.3. Effect of Vitamin D-Enriched Cheese Consumption on Bone Metabolism Biomarkers

Table 2 presents the GEE regression coefficients by vitamin D status of the under-study population of postmenopausal women. Women in the IG that consumed the vitamin D-enriched cheese and were vitamin D-insufficient women (25(OH)D < 50 nmol/L) at baseline were found to have higher 25(OH)D levels (Beta 13.8 ± 1.28; *p* < 0.001) and lower levels of the bone resorption marker TRAP-5b (Beta −0.20 ± 0.12; *p* = 0.050) at follow-up, compared with the CG that consumed the non-enriched cheese. Furthermore, women in the IG that were vitamin D-sufficient (25(OH)D ≥ 50 nmol/L) at baseline were found to have significantly higher levels of the bone formation marker P1NP (Beta 6.12 ± 2.42; *p* = 0.011) compared with women in the CG. No other significant between-group differences were observed, either in vitamin D-insufficient or in vitamin D-sufficient women.

Table 2. Effect of the intervention with vitamin D-enriched cheese compared with the control group (independent variable) on serum concentrations of calciotropic hormones and bone resorption markers at follow-up (dependent variables), by vitamin D status at baseline.

| | Vitamin D Status | | | | | |
| | 25(OH)D < 50 nmol/L (n = 48) | | | 25(OH)D ≥ 50 nmol/L (n = 31) | | |
Dependent Variables	Beta	SE	*p*-Value	Beta	SE	*p*-Value
Calciotropic hormones						
25(OH)D (nmol/L)	13.8	1.28	<0.001	4.48	2.01	0.126
PTH (pg/mL)	−0.91	0.95	0.067	1.15	1.09	0.108
Bone formation markers						
Osteocalcin (ng/mL)	0.12	0.81	0.883	1.54	1.11	0.166
P1NP (ng/mL)	−0.26	2.13	0.904	6.12	2.42	0.011
Bone resorption markers						
TRAP-5b (U/L)	−0.20	0.12	0.050	0.12	0.13	0.348

p-values obtained by GEE corrected for baseline values and because of significant interactions shown per group.

Table 3 shows the GEE regression coefficients by vitamin D status and years since menopause. In women that were vitamin D-insufficient at baseline and were in menopause for less than 5 years, consumption of vitamin D-enriched cheese was found to be associated with higher 25(OH)D levels (Beta 23.4 ± 6.39; *p* = 0.022) and lower levels of serum PTH (Beta −0.63 ± 1.11; *p* < 0.001) and TRAP-5b (Beta −0.65 ± 0.23; *p* = 0.004) at follow-up, compared with the CG that consumed the non-enriched cheese. Women that were vitamin D-insufficient at baseline and for more than 5 years in menopause were also found to have higher concentrations of 25(OH)D (Beta 13.4 ± 1.35; *p* < 0.001) at follow-up, compared with the CG, but there were no other significant associations observed with serum PTH and bone resorption markers. In women that were vitamin D-sufficient at baseline and less than 5 years in menopause, consumption of the vitamin D-enriched cheese was positively associated with the serum concentrations of osteocalcin (Beta 12.6 ± 1.46; *p* < 0.001), P1NP (Beta 6.24 ± 2.35; *p* = 0.008) and TRAP-5b (Beta 0.39 ± 0.11; *p* < 0.001). A positive association was also observed with P1NP concentrations (Beta 6.24 ± 2.35; *p* = 0.008) in vitamin D-sufficient women with more than 5 years since their menopause.

Table 3. Effect of the intervention with vitamin D-enriched cheese compared with the control group (independent variable) on serum concentrations of calciotropic hormones and bone resorption markers at follow-up (dependent variables), by vitamin D status at baseline and years since menopause.

	Vitamin D Status											
	25(OH)D < 50 nmol/L						25(OH)D ≥ 50 nmol/L					
	Years since Menopause						Years since Menopause					
	≤5 Years (n = 6)			>5 Years (n = 42)			≤5 Years (n = 3)			>5 Years (n = 28)		
Dependent Variables	Beta	SE	*p*-Value	Beta	SE	*p*-Value	Beta	SE	*p*-Value	Beta	SE	*p*-Value
Calciotropic hormones												
25(OH)D (nmol/L)	23.4	6.39	0.022	13.4	1.35	<0.001	6.50	11.4	0.669	3.14	2.28	0.182
PTH (pg/mL)	−0.63	1.11	<0.001	−0.95	1.05	0.301	−0.86	1.11	0.165	1.16	1.10	0.110

Table 3. *Cont.*

Dependent Variables	Vitamin D Status											
	25(OH)D < 50 nmol/L						25(OH)D $\geq$ 50 nmol/L					
	Years since Menopause						Years since Menopause					
	$\leq$5 Years (n = 6)			>5 Years (n = 42)			$\leq$5 Years (n = 3)			>5 Years (n = 28)		
	Beta	SE	*p*-Value	Beta	SE	*p*-Value	Beta	SE	*p*-Value	Beta	SE	*p*-Value
Bone formation markers												
Osteocalcin (ng/mL)	0.04	0.81	0.959	0.04	0.81	0.959	12.61	1.46	<0.001	0.55	1.04	0.598
P1NP (ng/mL)	−0.05	2.03	0.981	−0.05	2.03	0.981	6.24	2.35	0.008	6.24	2.35	0.008
Bone resorption markers												
TRAP-5b (U/L)	−0.65	0.23	0.004	−0.15	0.11	0.158	0.39	0.11	<0.001	0.04	0.13	0.750

p-values obtained by GEE corrected for baseline values and because of significant interactions shown per group.

4. Discussion

The present single-blinded, randomised, controlled trial examined the effect of vitamin D-enriched cheese consumption that provided 5.7 mcg of vitamin D_3 per day on the serum concentrations of certain calciotropic hormones and bone remodeling biomarkers on postmenopausal women living in Greece. The intervention was implemented from January to March, which is a typical winter period in Greece, and resulted in significant drops of serum 25(OH)D levels in the wider population due to limited sunlight exposure [27,28]. As previously reported [21], consumption of the vitamin D_3-enriched Gouda-type cheese significantly increased serum 25(OH)D concentrations by an average of 6.4 nmol/L from baseline in the examined postmenopausal women.

The present study is adding to these previous findings, reporting significant effects of the implemented vitamin D treatment on the examined biomarkers of bone metabolism and in women with different vitamin D and menopausal status. In this regard, in women that were vitamin D-insufficient at baseline and were allocated to the treatment arm that consumed the vitamin D-enriched cheese, 25(OH)D levels at follow-up were higher by an average of 23.4 and 13.4 nmol/L in those being at menopause for less and more than 5 years, respectively, compared with those consuming the non-enriched cheese. The increases reported by the present study in 25(OH)D levels are much higher than those reported by other similar studies on vitamin D-insufficient women. Bonjour et al. [29] observed a significant increase in serum 25(OH)D concentration of 9 nmol/L in postmenopausal women (mean age 57.1 ± 3.9 years), which were vitamin D-insufficient at study entry, following consumption of vitamin D-enriched cheese providing a daily dose of 2.5 mcg of vitamin D. The much higher 25(OH)D concentrations observed in the present study in early postmenopausal women (i.e., those being $\leq$5 years at menopause) consuming the vitamin D-enriched cheese, may also explain the lower PTH levels that were observed in the IG at follow-up, compared with the CG. These findings reflect the well-documented inverse relationship between 25(OH)D and PTH levels [30,31] and indicate that even a moderate supplemented daily dose of 5.7 mcg of vitamin D is adequate to inhibit the increase in PTH in blood among vitamin D-insufficient individuals.

The lowering effect of treatment with the vitamin D-enriched cheese on the serum PTH levels of the vitamin D-insufficient early postmenopausal women, could have also resulted to a lower rate of bone resorption, as evidenced by the lower levels of TRAP-5b at follow-up observed in the IG compared with their counterparts in the CG that consumed the non-enriched cheese. Other dietary intervention trials examining the effects of vitamin D-enriched dairy products on biomarkers of bone metabolism have also reported a suppression of bone resorption, mainly through a decrease in TRAP-5b concentrations [5,29,32,33]. TRAP-5b seems to have an advantage against other bone resorption biomarkers, since it is a lysosomal enzyme specific only to osteoclast activity [4], having a much lower day-by-day and within-subject variability as compared with urinary and other serum telopeptide molecules [34]. The above is of particular importance, particularly in trials with smaller sample sizes and thereby lower statistical power, as with the present study. In addition, TRAP-5b has been also reported to have a higher specificity and sensitivity in comparison with other bone resorption biomarkers [35], although it should be noted that CTX-1 is the recommended bone resorption biomarker by the International Osteoporosis Foundation and the International Federation of Clinical Chemistry and Laboratory Medicine [36]. Regarding vitamin D-insufficient women that were at menopause for more than 5 years, although the average serum concentration of 25(OH)D at follow-up was higher in the study group that consumed the vitamin D-enriched cheese, there was no significant effect observed on either PTH or bone resorption biomarkers markers. This probably indicates that the need of an intervention is more pronounced in vitamin D-insufficient women that are in their early postmenopausal phase, for whom the rate of bone turnover is accelerated [37].

In vitamin D-sufficient women, consumption of the vitamin D-enriched cheese had no effect on their vitamin D status, as there were no differences observed in 25(OH)D levels at follow-up between women allocated to the two treatment arms. However, as indicated by the significant positive beta coefficients reported in Table 2, early postmenopausal women (i.e., <5 years since menopause) with sufficient vitamin D levels (i.e., 25(OH)D > 50 nmol/L) in the IG were found to have higher serum levels of both bone formation (i.e., osteocalcin and P1NP) and bone resorption (i.e., TRAP-5b) markers compared with the CG, thus indicating an increased level of bone turnover. Increased rate of bone turnover has been associated with bone loss and fracture risk, particularly in individuals for whom vitamin D insufficiency coexists with hyperparathyroidism and low dietary calcium intake [38,39]. However, in the present study the increases observed in the levels of bone formation and bone resorption markers did not coincide with any of the above, since women that consumed the vitamin D-enriched cheese had sufficient levels of 25(OH)D and did not show any increase in their PTH levels, while their dietary calcium intake remained adequate throughout the study at levels higher than 900 mg per day [21]. In this regard, the observed acceleration in the rate of bone turnover could be attributed to other aetiological factors, with oestrogen deficiency occurring after menopause being one of them. It is known that the halt in oestrogen occurring after menopause in women is associated with increased bone turnover [40], which leads to a 9–10% decrease in bone mineral density (BMD), especially during the first 3 to 5 years after menopause [10]. In this regard, the lack of oestrogens might also negatively influence vitamin D-regulated processes, such as intestinal calcium absorption [41]. The above implies that the bone-enhancing effect of vitamin D-enriched cheese during menopause transition requires further study.

Cheese is an extremely good food source of both calcium and protein. In the present study, the addition of cheese in participants' diets increased both protein and calcium intake in both treatment arms, while those consuming the enriched cheese also increased their vitamin D intake. Based on the current knowledge, there is a synergy between vitamin D and calcium in terms of reducing the rate of bone resorption and turnover, at least in women with a low intake of calcium or vitamin D status [42]. Furthermore, high dietary protein intake stimulates the production of the anabolic hormone, insulin-like growth factor-I (IGF-I), which is important in bone formation, and has also been shown to have a mediating role in vitamin D metabolism. In this regard, higher circulating IGF-I levels have been reported to enhance renal production of 1,25-dihydroxyvitamin D, which also stimulates bone formation [42]. The synergy between dietary protein intake and vitamin D was also confirmed in a retrospective analysis with the use of data from a 3-year RCT examining calcium and vitamin D supplementation on 342 healthy people (≥65 years old) [43]. Those participants who completed the trial and were categorised in the highest tertile of dietary protein intake had an additional synergistic effect of higher protein, vitamin D and calcium intake on BMD at the femoral neck and total body [43]. In the present study, the interaction between dietary protein, calcium and higher baseline serum 25(OH)D concentrations in vitamin D-sufficient women may explain the increase in bone formation due to the vitamin D-enriched cheese in the vitamin D-sufficient women, while not showing an increase in 25(OH)D. In vitamin D-sufficient women who were still in menopause transition, oestrogen levels might have interfered with treatment, leading to an increase in both bone formation and bone resorption. In this context, the final effect of vitamin D-enriched cheese consumption on bone metabolism, and subsequently on BMD in vitamin D-sufficient women, will ultimately depend on the interactions between protein and calcium intake, vitamin D and PTH levels, as well as the decrease in oestrogen levels occurring after menopause. However, there is still a need for further study to confirm all aforementioned interactions on postmenopausal women's bone metabolism.

The findings of the present study should be interpreted in light of its strengths and limitations. The main strengths of the study are its randomised placebo-controlled design and the focus on vitamin D-insufficient women, who are most in need of a targeted intervention with vitamin D supplementation. Other important strengths include the detailed study protocols and procedures, which were tightly followed to assure the correct implementation of the intervention, the independent verification of the vitamin D content of foods and the high adherence of participants to treatment, as also confirmed by the procedures followed to record cheese consumption. Limitations of the present study are the relatively small sample size, especially of women that were up to 5 years post-menopause. In addition, although the study was adequately powered for the examination of the changes in its primary outcome, this might not be the case for the secondary outcomes examined in the present study, i.e., the serum levels of PTH and bone remodeling biomarkers. Another limitation of the current study was also its single-blinded design (i.e., blinded only to study participants). However, although those research team members that distributed the two different types of cheese to study participants were not blinded to treatment, the researchers undertaking, and subsequently reporting, the biochemical outcome measures were masked to all participants' allocation scheme. Furthermore, the 2 h waiting period might have been too long and may have resulted in the degradation of some sensitive biomarkers, such as osteocalcin. However, based on the standardised protocol followed in the clinical chemistry laboratory, where all biochemical analyses took place, the 2 h waiting period was an acceptable time for allowing the blood to adequately clot and avoiding haemolysis during centrifugation, while it only had minimal or no effect on most of the examined bone remodeling biomarkers. Lastly, although the biological variability of the examined bone remodelling biomarkers might have a confounding effect on the study findings, the methods followed in the present study for collecting, processing and analysing blood to determine the concentrations of bone remodeling biomarkers in serum were standardised, and were the same for all study participants and in both time points of data collection. In this regard, any biological variability produced a systematic error that was equally affecting subjects in both the intervention and the control group.

5. Conclusions

In conclusion, the present dietary intervention with Gouda-type cheese enriched with vitamin D significantly increased serum 25(OH)D concentrations, prevented PTH increase and reduced bone resorption in vitamin D-insufficient early postmenopausal women. In vitamin D-sufficient women, an interaction between the extra protein and calcium from the cheese combined with the higher vitamin D status might explain the increase in the concentration of bone formation markers. This increase in bone formation seemed to coincide with an increase in bone resorption and consequently turnover in women that were less than 5 years at menopause, which might be due to interference with declining oestrogen levels. As reduction in bone resorption observed in vitamin D-insufficient women concurs with increased 25(OH)D concentrations, the findings of the present study could reflect a novel food-based approach in improving vitamin D status in this susceptible population group and in producing some favourable changes on bone metabolism that may be protective against bone loss that occurs post-menopause. However, the interaction between the consumption of vitamin D-enriched cheese with serum 25(OH)D concentrations and oestrogen insufficiency in vitamin D-sufficient postmenopausal women requires further investigation.

Author Contributions: Conceptualisation, E.G.v.d.H. and Y.M.; Data curation, G.M. and C.M.; Formal analysis, G.M. and C.M.; Methodology, E.G.v.d.H. and Y.M.; Supervision, Y.M.; Writing—original draft, G.M., E.G.v.d.H., C.M. and Y.M.; Writing—review and editing, G.M., E.G.v.d.H., C.M. and Y.M. All authors have read and agreed to the published version of the manuscript.

Funding: This work was supported by funding received from the European Commission under its Seventh Framework Programme (FP7) ODIN (grant agreement No. 613977) and by grant received by FrieslandCampina BV for the biochemical analyses of PTH, OC, P1NP and TRAP-5b. All cheese products were also produced and provided by FrieslandCampina BV.

Institutional Review Board Statement: The study was conducted according to the guidelines of the Declaration of Helsinki, and approved by the Ethics Committee of Harokopio University (protocol code 43/23-07-2014).

Informed Consent Statement: Informed consent was obtained from all subjects involved in the study.

Acknowledgments: The authors would like to thank all study group members that contributed to the study execution: Eirini Efstathopoulou, Eftychia Apostolidou, Antigoni Tsiafitsa, Athanasios Douligeris, Konstantina Maragkopoulou, Paraskeui-Evita Siatitsa, Efstathoula Argiri and Chrysanthe Papafotiou, as well as all study participants. Support of M. Roordink for the development and production of vitamin D-enriched cheese and of Jette Jacobs for the measurement of vitamin D content in the cheese is also greatly acknowledged. Finally, special thanks go to Kevin D. Cashman and Mairead E. Kiely, who were the coordinators of the ODIN EU-funded project, under which this study was conducted.

Conflicts of Interest: E.G.v.d.H. was at the time of the study employee at FrieslandCampina. None of the other authors have any potential conflict of interest to declare. Any opinions, findings, conclusions, or recommendations expressed in the current study are those of the authors and do not necessarily reflect the views of FrieslandCampina.

References

1. Ross, A.C. The 2011 report on dietary reference intakes for calcium and vitamin D. *Public Health Nutr.* **2011**, *14*, 938–939. [CrossRef]
2. Holick, M.F. The vitamin D deficiency pandemic: Approaches for diagnosis, treatment and prevention. *Rev. Endocr. Metab. Disord.* **2017**, *18*, 153–165. [CrossRef]
3. Ebeling, P.R. Vitamin D and bone health: Epidemiologic studies. *BoneKEy Rep.* **2014**, *3*, 511. [CrossRef] [PubMed]
4. Heaney, R.P. The bone remodeling transient: Interpreting interventions involving bone-related nutrients. *Nutr. Rev.* **2001**, *59*, 327–334. [CrossRef] [PubMed]
5. Heaney, R.P.; McCarron, D.A.; Dawson-Hughes, B.; Oparil, S.; Berga, S.L.; Stern, J.S.; Barr, S.I.; Rosen, C.J. Dietary changes favorably affect bone remodeling in older adults. *J. Am. Diet. Assoc.* **1999**, *99*, 1228–1233. [CrossRef]
6. Nishizawa, Y.; Ohta, H.; Miura, M.; Inaba, M.; Ichimura, S.; Shiraki, M.; Takada, J.; Chaki, O.; Hagino, H.; Fujiwara, S.; et al. Guidelines for the use of bone metabolic markers in the diagnosis and treatment of osteoporosis (2012 edition). *J. Bone Miner. Metab.* **2013**, *31*, 1–15. [CrossRef]
7. Lombardi, G.; Perego, S.; Luzi, L.; Banfi, G. A four-season molecule: Osteocalcin. Updates in its physiological roles. *Endocrine* **2015**, *48*, 394–404. [CrossRef]
8. Krege, J.H.; Lane, N.E.; Harris, J.M.; Miller, P.D. PINP as a biological response marker during teriparatide treatment for osteoporosis. *Osteoporos. Int.* **2014**, *25*, 2159–2171. [CrossRef]
9. Cauley, J.A. Bone health after menopause. *Curr. Opin. Endocrinol. Diabetes Obes.* **2015**, *22*, 490–494. [CrossRef] [PubMed]
10. Greendale, G.A.; Sowers, M.; Han, W.; Huang, M.H.; Finkelstein, J.S.; Crandall, C.J.; Lee, J.S.; Karlamangla, A.S. Bone mineral density loss in relation to the final menstrual period in a multiethnic cohort: Results from the Study of Women's Health Across the Nation (SWAN). *J. Bone Miner. Res.* **2012**, *27*, 111–118. [CrossRef]
11. Rizzoli, R.; Stevenson, J.C.; Bauer, J.M.; van Loon, L.J.; Walrand, S.; Kanis, J.A.; Cooper, C.; Brandi, M.L.; Diez-Perez, A.; Reginster, J.Y.; et al. The role of dietary protein and vitamin D in maintaining musculoskeletal health in postmenopausal women: A consensus statement from the European Society for Clinical and Economic Aspects of Osteoporosis and Osteoarthritis (ESCEO). *Maturitas* **2014**, *79*, 122–132. [CrossRef]
12. Nieves, J.W. Osteoporosis: The role of micronutrients. *Am. J. Clin. Nutr.* **2005**, *81*, 1232S–1239S. [CrossRef]

13. Avenell, A.; Mak, J.C.; O'Connell, D. Vitamin D and vitamin D analogues for preventing fractures in post-menopausal women and older men. *Cochrane Database Syst. Rev.* **2014**. [CrossRef] [PubMed]
14. Lips, P.; Gielen, E.; van Schoor, N.M. Vitamin D supplements with or without calcium to prevent fractures. *BoneKEy Rep.* **2014**, *3*, 512. [CrossRef] [PubMed]
15. Baeksgaard, L.; Andersen, K.P.; Hyldstrup, L. Calcium and vitamin D supplementation increases spinal BMD in healthy, postmenopausal women. *Osteoporos. Int.* **1998**, *8*, 255–260. [CrossRef] [PubMed]
16. Dawson-Hughes, B.; Dallal, G.E.; Krall, E.A.; Sadowski, L.; Sahyoun, N.; Tannenbaum, S. A controlled trial of the effect of calcium supplementation on bone density in postmenopausal women. *N. Engl. J. Med.* **1990**, *323*, 878–883. [CrossRef] [PubMed]
17. Reid, I.R.; Ames, R.W.; Evans, M.C.; Gamble, G.D.; Sharpe, S.J. Long-term effects of calcium supplementation on bone loss and fractures in postmenopausal women: A randomized controlled trial. *Am. J. Med.* **1995**, *98*, 331–335. [CrossRef]
18. Vieth, R.; Chan, P.C.; MacFarlane, G.D. Efficacy and safety of vitamin D3 intake exceeding the lowest observed adverse effect level. *Am. J. Clin. Nutr.* **2001**, *73*, 288–294. [CrossRef] [PubMed]
19. Weisman, S.M.; Matkovic, V. Potential use of biochemical markers of bone turnover for assessing the effect of calcium supplementation and predicting fracture risk. *Clin. Ther.* **2005**, *27*, 299–308. [CrossRef] [PubMed]
20. Whiting, S.J.; Kohrt, W.M.; Warren, M.P.; Kraenzlin, M.I.; Bonjour, J.P. Food fortification for bone health in adulthood: A scoping review. *Eur. J. Clin. Nutr.* **2016**, *70*, 1099–1105. [CrossRef] [PubMed]
21. Manios, Y.; Moschonis, G.; Mavrogianni, C.; van den Heuvel, E.; Singh-Povel, C.M.; Kiely, M.; Cashman, K.D. Reduced-fat Gouda-type cheese enriched with vitamin D3 effectively prevents vitamin D deficiency during winter months in postmenopausal women in Greece. *Eur. J. Nutr.* **2017**, *56*, 2367–2377. [CrossRef]
22. Burild, A.; Frandsen, H.L.; Poulsen, M.; Jakobsen, J. Quantification of physiological levels of vitamin D(3) and 25-hydroxyvitamin D(3) in porcine fat and liver in subgram sample sizes. *J. Sep. Sci.* **2014**, *37*, 2659–2663. [CrossRef] [PubMed]
23. Ezzati, M.; Bentham, J.; Di Cesare, M.; Bilano, V.; Bixby, H.; Zhou, B.; Stevens, G.A.; Riley, L.M.; Taddei, C.; Hajifathalian, K.; et al. Worldwide trends in body-mass index, underweight, overweight, and obesity from 1975 to 2016: A pooled analysis of 2416 population-based measurement studies in 128.9 million children, adolescents, and adults. *Lancet* **2017**, *390*, 2627–2642. [CrossRef]
24. Cashman, K.D.; Kiely, M.; Kinsella, M.; Durazo-Arvizu, R.A.; Tian, L.; Zhang, Y.; Lucey, A.; Flynn, A.; Gibney, M.J.; Vesper, H.W.; et al. Evaluation of Vitamin D Standardization Program protocols for standardizing serum 25-hydroxyvitamin D data: A case study of the program's potential for national nutrition and health surveys. *Am. J. Clin. Nutr.* **2013**, *97*, 1235–1242. [CrossRef]
25. Institute of Medicine Food and Nutrition Board. *Dietary Reference Intakes for Calcium and Vitamin D*; National Academy Press: Washington, DC, USA, 2011.
26. Burger, H.G.; Hale, G.E.; Dennerstein, L.; Robertson, D.M. Cycle and hormone changes during perimenopause: The key role of ovarian function. *Menopause* **2008**, *15*, 603–612. [CrossRef]
27. Manios, Y.; Moschonis, G.; Hulshof, T.; Bourhis, A.S.; Hull, G.L.J.; Dowling, K.G.; Kiely, M.E.; Cashman, K.D. Prevalence of vitamin D deficiency and insufficiency among schoolchildren in Greece: The role of sex, degree of urbanisation and seasonality. *Br. J. Nutr.* **2017**, *118*, 550–558. [CrossRef]
28. Manios, Y.; Moschonis, G.; Lambrinou, C.P.; Tsoutsoulopoulou, K.; Binou, P.; Karachaliou, A.; Breidenassel, C.; Gonzalez-Gross, M.; Kiely, M.; Cashman, K.D. A systematic review of vitamin D status in southern European countries. *Eur. J. Nutr.* **2018**, *57*, 2001–2036. [CrossRef] [PubMed]
29. Bonjour, J.P.; Benoit, V.; Pourchaire, O.; Ferry, M.; Rousseau, B.; Souberbielle, J.C. Inhibition of markers of bone resorption by consumption of vitamin D and calcium-fortified soft plain cheese by institutionalised elderly women. *Br. J. Nutr.* **2009**, *102*, 962–966. [CrossRef]
30. Lips, P. Vitamin D deficiency and secondary hyperparathyroidism in the elderly: Consequences for bone loss and fractures and therapeutic implications. *Endocr. Rev.* **2001**, *22*, 477–501. [CrossRef]
31. Chapuy, M.C.; Preziosi, P.; Maamer, M.; Arnaud, S.; Galan, P.; Hercberg, S.; Meunier, P.J. Prevalence of vitamin D insufficiency in an adult normal population. *Osteoporos. Int.* **1997**, *7*, 439–443. [CrossRef] [PubMed]
32. Bonjour, J.P.; Benoit, V.; Pourchaire, O.; Rousseau, B.; Souberbielle, J.C. Nutritional approach for inhibiting bone resorption in institutionalized elderly women with vitamin D insufficiency and high prevalence of fracture. *J. Nutr. Health Aging* **2011**, *15*, 404–409. [CrossRef]
33. Bonjour, J.P.; Benoit, V.; Rousseau, B.; Souberbielle, J.C. Consumption of vitamin D-and calcium-fortified soft white cheese lowers the biochemical marker of bone resorption TRAP 5b in postmenopausal women at moderate risk of osteoporosis fracture. *J. Nutr.* **2012**, *142*, 698–703. [CrossRef]
34. Seibel, M.J.; Robins, S.P.; Bilezikian, J.P. *Dynamics of Bone and Cartilage Metabolism*, 2nd ed.; Academic Press: San Diego, CA, USA, 2006; pp. 529–540.
35. Kuo, T.R.; Chen, C.H. Bone biomarker for the clinical assessment of osteoporosis: Recent developments and future perspectives. *Biomark Res.* **2017**, *5*, 18. [CrossRef]
36. Vasikaran, S.; Cooper, C.; Eastell, R.; Griesmacher, A.; Morris, H.A.; Trenti, T.; Kanis, J.A. International Osteoporosis Foundation and International Federation of Clinical Chemistry and Laboratory Medicine position on bone marker standards in osteoporosis. *Clin. Chem. Lab. Med.* **2011**, *49*, 1271–1274. [CrossRef] [PubMed]
37. Riggs, B.L. Endocrine causes of age-related bone loss and osteoporosis. *Novart. Fdn. Symp.* **2002**, *242*, 247–259.

38. Iki, M.; Morita, A.; Ikeda, Y.; Sato, Y.; Akiba, T.; Matsumoto, T.; Nishino, H.; Kagamimori, S.; Kagawa, Y.; Yoneshima, H.; et al. Biochemical markers of bone turnover predict bone loss in perimenopausal women but not in postmenopausal women-the Japanese Population-based Osteoporosis (JPOS) Cohort Study. *Osteoporos. Int.* **2006**, *17*, 1086–1095. [CrossRef] [PubMed]
39. Lofman, O.; Magnusson, P.; Toss, G.; Larsson, L. Common biochemical markers of bone turnover predict future bone loss: A 5-year follow-up study. *Clin. Chim. Acta* **2005**, *356*, 67–75. [CrossRef] [PubMed]
40. Perrien, D.S.; Achenbach, S.J.; Bledsoe, S.E.; Walser, B.; Suva, L.J.; Khosla, S.; Gaddy, D. Bone turnover across the menopause transition: Correlations with inhibins and follicle-stimulating hormone. *J. Clin. Endocrinol. Metab.* **2006**, *91*, 1848–1854. [CrossRef] [PubMed]
41. Colin, E.M.; Van Den Bemd, G.J.; Van Aken, M.; Christakos, S.; De Jonge, H.R.; Deluca, H.F.; Prahl, J.M.; Birkenhager, J.C.; Buurman, C.J.; Pols, H.A.; et al. Evidence for involvement of 17beta-estradiol in intestinal calcium absorption independent of 1,25-dihydroxyvitamin D3 level in the Rat. *J. Bone Miner. Res.* **1999**, *14*, 57–64. [CrossRef]
42. Cashman, K.D.; van den Heuvel, E.G.; Schoemaker, R.J.; Preveraud, D.P.; Macdonald, H.M.; Arcot, J. 25-Hydroxyvitamin D as a Biomarker of Vitamin D Status and Its Modeling to Inform Strategies for Prevention of Vitamin D Deficiency within the Population. *Adv. Nutr.* **2017**, *8*, 947–957. [CrossRef]
43. Dawson-Hughes, B.; Harris, S.S. Calcium intake influences the association of protein intake with rates of bone loss in elderly men and women. *Am. J. Clin. Nutr.* **2002**, *75*, 773–779. [CrossRef]

Article

Plasma Fatty Acids and Quantitative Ultrasound, DXA and pQCT Derived Parameters in Postmenopausal Spanish Women

Raúl Roncero-Martín [1,†], Ignacio Aliaga [2,†], Jose M. Moran [1,*], Luis M. Puerto-Parejo [1], Purificación Rey-Sánchez [1], María de la Luz Canal-Macías [1], Antonio Sánchez-Fernández [3], Juan D. Pedrera-Zamorano [1], Fidel López-Espuela [1], Vicente Vera [2], Purificación Cerrato-Carretero [1] and Jesús M. Lavado-García [1]

[1] Metabolic Bone Diseases Research Group, Nursing and Occupational Therapy College, University of Extremadura, 10003 Caceres, Spain; rronmar@unex.es (R.R.-M.); lmpuerto@unex.es (L.M.P.-P.); prey@unex.es (P.R.-S.); luzcanal@unex.es (M.d.l.L.C.-M.); jpedrera@unex.es (J.D.P.-Z.); fidellopez@unex.es (F.L.-E.); puricc@yahoo.es (P.C.-C.); jmlavado@unex.es (J.M.L.-G.)
[2] Department of Conservative Dentistry and Prostheses, Faculty of Dentistry, Complutense University, 28040 Madrid, Spain; i.aliaga@pdi.ucm.es (I.A.); vveragon@ucm.es (V.V.)
[3] Servicio de Tocoginecología, Complejo Hospitalario de Cáceres, 10004 Caceres, Spain; gineantonio@gmail.com
* Correspondence: jmmorang@unex.es; Tel.: + 34-927-257450
† These authors contributed equally to this work.

Citation: Roncero-Martín, R.; Aliaga, I.; Moran, J.M.; Puerto-Parejo, L.M.; Rey-Sánchez, P.; de la Luz Canal-Macías, M.; Sánchez-Fernández, A.; Pedrera-Zamorano, J.D.; López-Espuela, F.; Vera, V.; et al. Plasma Fatty Acids and Quantitative Ultrasound, DXA and pQCT Derived Parameters in Postmenopausal Spanish Women. *Nutrients* **2021**, *13*, 1454. https://doi.org/10.3390/nu13051454

Academic Editor: Andrew J. Sinclair

Received: 8 March 2021
Accepted: 21 April 2021
Published: 25 April 2021

Publisher's Note: MDPI stays neutral with regard to jurisdictional claims in published maps and institutional affiliations.

Abstract: Different factors may contribute to the development of osteopenia or osteoporosis. Fatty acids are key nutrients for health, and a number of studies have reported an association between bone mineral density (BMD) and fatty acid intake. We aimed to investigate the relationships between serum levels of different fatty acids and bone parameters determined by quantitative bone ultrasound (QUS), peripheral quantitative computed tomography (pQCT), and dual-energy X-ray absorptiometry (DXA) in a sample of Spanish postmenopausal women. We enrolled a total of 301 postmenopausal women (median age 59 years; interquartile range (IQR) 7) in this study. All participants underwent full densitometric screening, including calcaneal quantitative ultrasound (QUS), peripheral quantitative computed tomography (pQCT), and dual-energy X-ray absorptiometry (DXA), as well as plasma fatty acid measurement. After adjustment for potential confounders, plasma n-3 polyunsaturated fatty acid (PUFA) levels correlated with BMD in the spine (r = 0.150; p = 0.014) and femoral neck (r = 0.143; p = 0.019). By multiple linear regression, an independent statistically significant positive relationship was observed between BMD in the spine and BMI (β = 0.288; p = 0.001) as well as total plasma n-3 PUFAs (β = 0.155; p = 0.009). The plasma n-3 PUFA level was also a significant and positive predictor of BMD at the femoral neck (β = 0.146; p = 0.009). Independent risk factors for low BMD (T-score ≤ 1) were determined by logistic regression analysis, and a relatively high level of plasma n-3 PUFAs (OR = 0.751; 95% CI 0.587-0.960, p = 0.022) was identified as a protective factor against low bone mass. In this single-center sample of Spanish postmenopausal women, we reported a significant positive and statistically independent association between BMD and plasma levels of n-3 PUFAs.

Keywords: fatty acids; n-3 PUFAs; bone mineral density; osteoporosis; postmenopause

1. Introduction

Several diverse factors may contribute to the development of osteopenia/osteoporosis. Some of the most important are a sedentary lifestyle, inadequate nutrient consumption, inflammation, and genetic factors. The essential fatty acids are nutrients of key importance for health. Previous reports regarding the influence of diet on bone health in healthy populations established a link from fatty acid intake to total bone mineral density (BMD) through a number of mechanisms that promote bone formation [1–3].

The long-chain n-3 polyunsaturated fatty acids (n-3 PUFAs) α-linolenic acid (ALA) is mainly consumed from food sources, such as various nuts and seeds (walnuts, flaxseeds, chia seeds) and vegetable oils (linseed oil, canola oil, soybean oil), while eicosapentaenoic acid (EPA) and docosahexaenoic acid (DHA) are key fatty acids that are found almost exclusively in marine foods such as fatty fish [4]. *n*-6 PUFAs (linoleic acid (LA) and arachidonic acid (AA)) and *n*-3 fatty acids are of specific interest because they contribute to the structure and function of the phospholipid bilayers that constitute cell membranes and because they are precursors of eicosanoids, comprising prostaglandins, leukotrienes, and thromboxanes, with hormone-like activities [5]. *n*-3 PUFAs have the potential to benefit the bones, as increased consumption rates appear to be linked with increased BMD [6–9]. Mechanisms postulated for the effect of *n*-3 PUFAs on bone health include an indirect effect based on enhancing intestinal calcium uptake, together with direct effects on bone turnover based on affecting the activity of osteoblasts and osteoclasts [10–13]. It has been suggested that the intake of polyunsaturated lipids may have an influence on bone mineral accrual and BMD and may even play a relevant role in the prevention of fragility fractures [3,14–16]. The strongest evidence comes from observational studies that have reported how total PUFA intake, particularly *n*-3 and *n*-6 PUFA intake, may increase BMD and even reduce fracture risk [3,17–20]. Nevertheless, the literature is very limited, particularly with regard to monounsaturated fatty acids (MUFAs) and saturated fatty acids (SFAs) and their association with BMD or the risk of fracture [21]. A limited number of studies have addressed the role of these fatty acids in particular, finding that monounsaturated fatty acid intake [14] may decrease total fracture risk.

However, the relationship between dietary PUFA consumption and hip fracture risk has shown controversial results in several observational studies, some with large sample sizes. The Nurses' Health Study (NHS) followed postmenopausal women for 24 years and showed a statistically significant decrease in age-adjusted hip fracture risk in women with higher total *n*-3 PUFA intake compared to those with lower intake [15]. In a further observational cohort of over 135,000 postmenopausal US women participating in the Women's Health Initiative (WHI), total (*n*-3 and *n*-6) PUFA intake was also assessed and was found to be linked to a reduction in overall fracture risk after an average follow-up of 7.8 years, suggesting that an increased intake of certain *n*-3 PUFAs such as ALA and the intake of total *n*-6 PUFAs would be related to a reduced risk of fragility fractures [17]. However, as it has been indicated, these results contradict other available findings regarding the intake of fatty acids and their association with fractures and BMD. In most of the studies demonstrating an association, the authors note that it is not possible to determine the causality of these associations due to the observational nature of the studies [9,18], potentially making the observed results a consequence of residual confounding or reverse causality, in which case the inferred causality would be spurious.

Serum fatty acids circulating levels are ultimately determined by dietary consumption and biosynthesis. For *n*-3 and *n*-6 PUFAs, the percentages of LA, AA, EPA and DHA in total plasma fatty acids, are acceptable marker of their habitual consumption, but the percentage of ALA is not [22–24]. Long-chain *n*-6 and *n*-3 PUFA plasma concentrations are not affected by the intake levels of their precursors, LA and ALA [25]. This may explain some of the controversial findings regarding the intake of these fatty acids and their relationship to bone health. In this study, we assessed the relationships between serum levels of different PUFAs (*n*-6 and *n*-3), MUFAs and SFAs with bone density determined by quantitative bone ultrasound (QUS), peripheral quantitative computed tomography (pQCT), and dual-energy X-ray absorptiometry (DXA) in a sample of postmenopausal Spanish women.

2. Materials and Methods

2.1. Subjects

Subjects: In this cross-sectional study, a total of 301 postmenopausal women (median age 59 years; interquartile range (IQR) 7) that resided in the Urban Health District of Cáceres, Extremadura, Spain were enrolled from January 2019 to December 2019 in the local area from both primary and specialty care settings. Participants were recruited by convenience sampling from clinics. The participants underwent full densitometric screening, including calcaneal quantitative ultrasound (QUS), peripheral quantitative computed tomography (pQCT), and dual-energy X-ray absorptiometry (DXA). To be eligible for this study, these women were all required to be community residents of white European descent and not to have been diagnosed with functional mental or physical disabilities by either their current primary care physician or a specialist actively participating in their medical care. They were not prescribed any type of medication that might disrupt calcium metabolism (e.g., corticosteroids, oral anticoagulants, antipsychotics, etc.) and had no conditions, including those presumed to be potentially involved in disorders of mineral metabolism (diabetes mellitus, liver disease, renal osteodystrophy, or parathyroid, thyroid, adrenal, or ovarian disease), that would interfere with calcium metabolism. The subjects were all active, although none of them played any sport professionally The Ethical Advisory Committee of the University of Extremadura endorsed this study (protocol code 84/2018 and date of approval 6 July 2018). All the participants gave written informed consent in accordance with the 1975 Declaration of Helsinki.

2.2. Anthropometry

Height was measured using a Harpenden stadiometer with a mandibular plane that was parallel to the ground, while weight measurements were obtained with a biomedical precision scale. Height was accurately assessed to the nearest centimeter, and weight was measured to the nearest 100 g. All these measurements were performed while the subjects were only lightly clothed and shoeless. Alcohol consumption was reported to be occasional and did not exceed 100 mL/day. Body mass index (BMI) was computed as weight in kilograms divided by the square of height in meters (kg/m^2).

2.3. Dietary Assessment

All women who participated in this study were provided with a 131-item comprehensive 7-day food frequency questionnaire. Food was quantified using a dietetic scale, measuring cups, and spoons. The questionnaire used was self-reported. The response rate was 91.3%. This questionnaire has been used extensively in the literature [26]. Nutrient and energy intake values were assessed according to the Spanish food composition database [27].

2.4. Quantitative Bone Ultrasound

Heel bone characteristics were evaluated with a GE Sahara (Hologic, Bedford, MA, USA) quantitative ultrasound (QUS) device. This apparatus acquires two key parameters: broadband ultrasound attenuation (BUA), expressed in dB/MHz, provides a measure of ultrasound decay with the incident frequency of the sound of the waves, and the speed of sound (SOS), expressed in meters per second, reflects the time necessary for the ultrasound waves to travel a given distance across the calcaneal bone.

2.5. Bone Densitometric Determinations

Peripheral quantitative computed tomography (pQCT) scans of the nondominant distal forearm were performed with a Stratec XCT-2000 scanner (Stratec Medizintechnik, Pforzheim, Germany). The equipment was positioned at the distal end of the nondominant forearm, and 4% of the forearm length was scanned. The XCT-2000 measurement data were processed with version 5.50 of the software supplied by the supplier. A pQCT scan provides a measure of volumetric bone mineral density and discriminates between trabecular bone and cortical bone.

Women in this study also underwent bone densitometry by dual-energy X-ray absorptiometry (DXA) of the lumbar spine (L2-L4, L2, L3, L4) and hip (left femoral neck and femoral trochanter) using a Norland XR-800 densitometer (Norland at Swissray, Fort Atkinson, WI, USA). All BMD measurements were given as the quantity of minerals divided by the scanned area (g/cm^2). Participants were classified into the osteopenia, osteoporosis or normal group on the basis of the T-score at either the femoral neck or the spine (L2-L4).

The coefficient of variation (CV%) was below 2% in all bone measurements. We scanned an anthropomorphic pQCT phantom in each session to guarantee quality.

2.6. Determination of Plasma Fatty Acids

Fasting blood samples were collected, and plasma was stored at $-80\ °C$ until further analysis. Lipids were extracted from plasma samples, and fatty acids were isolated and separated by gas chromatography with flame ionization detection (GC-FID). GC–FID analysis was conducted using a Bruker Scion 456 GC equipped with a flame ionization detector (FID) and a DB-225 ms (Agilent Technologies) capillary column (30 m $\times$ 0.25 mm i.d., 0.25 µm film thickness), high-purity helium as the carrier gas, and a 0.2 µL injection volume, using a split/splitless capillary injection system at 220 °C with a split ratio of 100:1. The temperature program was as follows: initial temperature 140 °C for 1 min, increase by 10 °C/min to 200 °C, hold for 10 min, increase by 5 °C/min to 220 °C, and hold for 30 min. The total analysis run time was 51 min. A total of 17 different fatty acids were assessed: C8:0 (caprylic acid), C10:0 (decanoic acid), C12:0 (lauric acid), C14:0 (myristic acid), C16:0 (palmitic acid), C18:0 (stearic acid), C20:0 (arachidic acid), C22:0 (docosanoic acid), C24:0 (lignoceric acid), C16:1 (palmitoleic acid), C18:1 cis (n9) (oleic acid), C22:1 (n9) (erucic acid), C18:3 (n3) (α-linolenic acid, or ALA), C20:5 (n3) (eicosapentaenoic acid, or EPA), C22:6 (n3) (docosahexaenoic acid, or DHA), C18:2 (n6) (linoleic acid, or LA) and C20:4 (n6) (arachidonic acid, or AA). The average coefficient of variation was $\sim$60%.

2.7. Statistical Analysis

The results are reported as medians with interquartile ranges and frequency counts with percentages, unless otherwise specified. Continuous variables were analyzed using the Kruskal–Wallis test, the Mann–Whitney U test, and the Quade test, while categorical variables were analyzed using the chi-square test or Fisher's exact test as appropriate. Pairwise comparisons between each independent group were analyzed by Dunn's test with Bonferroni adjustment. The bivariate correlation analysis was carried out by calculating Spearman's correlation coefficient and then adjusting (partial correlations) for potential confounding factors. Given that several variables included in this study did not fulfill the normality criteria (by the Kolmogorov–Smirnov test) for the multiple linear regression study, a two-step method was used to normalize the data prior to statistical analyses [28]. The variables included in the modeling were age (years), time since menopause (years), BMI (kg/m^2), vitamin D (mcg/day), calcium intake (mg/day), energy (kcal/day), total plasma *n*-3 PUFAs, total plasma *n*-6 PUFAs, plasma *n*-6/*n*-3 ratio, total plasma SFAs, total plasma MUFAs, and total plasma PUFAs. Logistic regression was used to assess the probability of patients reaching an at-risk state for low bone mass, defined as a T-score < -1. The logistic regression included age (years), time since menopause (years), BMI (kg/m^2), vitamin D (mcg/day), calcium intake (mg/day), and energy intake (kcal/day).

For all statistical analyses, a two-sided p value ≤ 0.05 was considered statistically significant. All analyses were performed with SPSS software (version 24.0, IBM Corp., Armonk, NY, USA).

3. Results

3.1. Descriptive Data

The anthropometric, biological and dietary characteristics studied in the group of women are shown in Table 1. Participants were grouped by diagnosis of osteopenia/osteoporosis based on WHO T-score criteria, and these groups were not significantly different in weight, age at menarche or years since menopause ($p > 0.05$). A significant difference was observed in BMI, which was significantly increased in the group of healthy women ($p < 0.05$). Overall, 80.7% ($n = 243$) of the participants were nonsmokers. No differences were observed in the intake of vitamin D, calcium, or energy according to the dietary questionnaire ($p > 0.05$).

Table 1. Anthropometric, biological, dietary and lifestyle characteristics in the study sample.

	Total Sample ($n = 301$) Median (IQR); n (%)	Normal ($n = 103$) Median (IQR); n (%)	Osteopenia ($n = 145$) Median (IQR); n (%)	Osteoporosis ($n = 53$) Median (IQR); n (%)	p-Value
Age, years	59 (7)	58 (7)	60 (7)	60 (6)	0.071
Menarche age, years	13 (2)	12 (3)	13 (1)	13 (1)	0.842
Years since menopause, years	9 (9)	9 (10)	8 (9)	10 (10)	0.277
Weight, kg	66.4 (15.6)	70.6 (15.2) (a,b)	65.5 (13.8) (c)	58.1 (13.6)	<0.001
Height, m	1.58 (0.07)	1.59 (0.07)	1.58 (0.07)	1.57 (0.07)	0.078
BMI (kg/m^2)	26.6 (5.6)	27.7 (5.9) (a,b)	26.6 (5.3) (c)	24.7 (4)	<0.001
BMI Classification					
Underweight (<18.5)	$n = 1$ (0.3%)	$n = 1$ (1%)	$n = 0$ (0%)	$n = 0$ (0%)	0.518
Normal weigth (18.5–24.9)	$n = 98$ (32.6%)	$n = 23$ (22.3%)(a,b)	$n = 46$ (31.7%) (c)	$n = 29$ (54.7%)	0.003
Overweight (25.0–29.9)	$n = 135$ (44.9%)	$n = 47$ (45.6%) (a,b)	$n = 73$ (50.3%) (c)	$n = 15$ (28.3%)	0.02
Obesity class I (30.0–34.9)	$n = 50$ (16.6%)	$n = 21$ (20.4%)	$n = 23$ (15.9%)	$n = 6$ (11.3%)	0.359
Obesity class II (35.0–39.9)	$n = 12$ (4%)	$n = 8$ (7.8%) (a,b)	$n = 2$ (1.4%) (c)	$n = 2$ (3.8%)	0.02
Obesity class III ($\geq$40)	$n = 5$ (1.7%)	$n = 3$ (2.9%)	$n = 1$ (0.7%)	$n = 1$ (1.9%)	0.327
Waist circumference, cm	87 (14)	91 (17) (a,b)	87 (13) (c)	82 (13)	<0.001
Hip, cm	104 (12)	107 (13) (a)	104 (11) (c)	100 (14)	<0.001
Gravidity	2 (1)	2 (1)	2 (1)	2 (1)	0.256
Parity	2 (1)	2 (0)	2 (1)	2 (1)	0.44
Smoker					
No	$n = 243$ (80.70%)	$n = 86$ (83.5%)	$n = 116$ (80%)	$n = 41$ (77.4%)	0.597
Yes	$n = 58$ (19.3%)	$n = 17$ (16.5%)	$n = 29$ (20%)	$n = 12$ (22.6%)	
Fish intake (servings/week)	4 (3)	3 (4)	4 (3)	4 (3)	0.783
Vitamin D (mcrg/day)	7.4 (8.23)	7.4 (8)	7.6 (7.83)	7.8 (12.26)	0.707
Calcium intake, mg/day	944 (663)	930 (516)	959 (652)	882 (821)	0.908
Energy, kcal/day	2099 (869)	2087 (883)	2048 (860)	2204 (874)	0.892

Between-group comparisons were performed using the Kruskal Wallis test or the Fisher exact test as appropriate. (a) Poshoc analysis by Dunn's test, $p < 0.05$ vs. osteoporosis group. (b) Poshoc analysis by Dunn's test, $p < 0.05$ vs. osteopenia group. (c) Poshoc analysis by Dunn's test, $p < 0.05$ vs. osteoporosis group.

The fatty acid profile of the study participants is shown in Table 2. A total of 17 different fatty acids were quantified. When subjects were categorized based on osteopenia/osteoporosis criteria, statistically significant differences were observed in the levels of C12:0 (lauric acid) ($p = 0.029$), C16:0 (palmitic acid) ($p = 0.018$), C24:0 (lignoceric acid) ($p = 0.043$), C18:1 cis (n9) (oleic acid) ($p = 0.013$), and total plasma MUFAs ($p = 0.016$). No statistically significant differences were observed in the levels of plasma total n-3 PUFAs, plasma total n-6 PUFAs, plasma n-6/n-3 ratio, plasma total SFAs, or total PUFAs ($p > 0.05$ in all cases). The statistically significant differences observed remained after adjusting for potential confounding factors in C16:0 palmitic acid ($p = 0.036$), C24:0 lignoceric acid ($p = 0.042$), and C18:1 cis (n9) oleic acid ($p = 0.045$).

Table 2. Plasma fatty acids profiles (%) in the studied sample.

	Total Sample (n = 330)	Normal (n = 103)	Osteopenia (n = 145)	Osteoporosis (n = 53)	p Value	Adjusted p Value *
	Median (IQR)	Median (IQR)	Median (IQR)	Median (IQR)		
C8:0 Caprylic acid	0.06 (0.09)	0.06 (0.07)	0.05 (0.09)	0.08 (0.07)	0.057	
C10:0 Decanoic acid	0 (0)	0 (0)	0 (0)	0 (0)	N/A	
C12:0 Lauric acid	0.07 (0.48)	0.06 (0.25) (a)	0.09 (0.55)	0.07 (0.08)	0.029	0.079
C14:0 Myristic acid	0.78 (0.28)	0.77 (0.26)	0.78 (0.31)	0.77 (0.25)	0.953	
C16:0 Palmitic acid	39.88 (3.69)	40.2 (3.36) (a)	39.28 (3.49) (b)	40.22 (4.39)	0.018	0.036
C18:0 Stearic acid	24.99 (5.04)	25.23 (3.86)	24.72 (5.93)	25.78 (3.53)	0.195	
C20:0 Arachidic acid	0.28 (1.08)	0.28 (0.89)	0.27 (1.11)	0.87 (1.21)	0.219	
C22:0 Docosanoic acid	0.34 (0.17)	0.33 (0.16)	0.35 (0.18)	0.33 (0.14)	0.796	
C24:0 Lignoceric acid	0.25 (0.15)	0.25 (0.14)	0.26 (0.17) (b)	0.22 (0.1)	0.043	0.042
C16:1 Palmitoleic acid	0.61 (0.39)	0.6 (0.39)	0.66 (0.4)	0.66 (0.38)	0.493	
C18:1 cis (n9) Oleic acid	10.16 (4.16)	9.75 (3.01) (a)	10.82 (4.47)	9.57 (3.85)	0.013	0.045
C22:1 (n9) Erucic acid	0 (0)	0 (0)	0 (0)	0 (0)	N/A	
C18:3 (n3) Linolenic acid (ALA)	0.31 (0.77)	0.33 (0.97)	0.3 (0.75)	0.3 (0.17)	0.309	
C20:5 (n3) Eicosapentenoic acid (EPA)	0.3 (0.27)	0.28 (0.19)	0.34 (0.29)	0.26 (0.33)	0.251	
C22:6 (n3) Docosahexenoic acid (DHA)	1.42 (0.92)	1.52 (0.85)	1.34 (0.97)	1.33 (0.87)	0.123	
C18:2 (n6) Linoleic acid (LA)	14.06 (4.12)	13.77 (3.98)	14.64 (4.43)	13.46 (3.4)	0.057	
C20:4 (n6) Arachidonic acid (AA)	5.04 (1.95)	5.18 (1.65)	5 (2.4)	4.75 (1.28)	0.150	
Plasma total n-3 PUFA	2.33 (1.53)	2.54 (1.29)	2.17 (1.78)	2.34 (1.35)	0.069	
Plasma total n-6 PUFA	18.92 (5.09)	18.85 (4.6)	19.11 (5.42)	18.64 (3.88)	0.247	
Plasma n-6/n-3 ratio	8.28 (5.85)	7.94 (4.57)	8.93 (8.14)	7.77 (6.58)	0.091	
Plasma total SFA	67.75 (7.45)	68.34 (6.93)	67.13 (8.58)	68.48 (5.22)	0.079	
Plasma total MUFA	10.9 (4.31)	10.42 (3.08) (a)	11.75 (4.75)	10.37 (3.65)	0.016	0.057
Plasma total PUFA	21.48 (4.43)	21.48 (3.78)	21.68 (5.05)	20.85 (3.76)	0.284	

Between-group comparisons were performed using the Kruskal Wallis test. (a) Poshoc analysis by Dunn's test, p < 0.05 vs. osteopenia group. (b) Poshoc analysis by Dunn's test, p < 0.05 vs. osteoporosis group * Adjusted by, age, years since menopause, BMI and vitamin D, calcium and energy intake (Quade's test).

3.2. Bone Parameters and Fatty Acid Plasma Levels

In order to explore the potential role of n-3 PUFAs in determining bone density, participants were classified based on the tertile of plasma n-3 PUFA levels (Table 3). This analysis revealed statistically significant differences in cortical volumetric density (p = 0.048) and BMD in the total spine (p = 0.011), L2 (p = 0.023), L3 (p = 0.033), and L4 (p = 0.004). The Z-score and T-score at the lumbar spine also showed statistically significant differences between the study groups (p < 0.05 in both cases). Regarding the hip, statistically significant differences were observed at the femoral neck in terms of the T-score (p = 0.042) as well as the Z-score (p = 0.035). We then proceeded to adjust this analysis for potential confounding factors present in the study sample, such as age; years since menopause; BMI; and intake of vitamin D, calcium, and energy. Differences observed in cortical volumetric BMD remained statistically significant (p = 0.013), as did those reported at the lumbar spine (p = 0.043) and L4 (p = 0.021). Previously reported differences in spine T-score and BMD at the femoral neck level were no longer statistically significant after adjustment (p > 0.05). Across all scores, the bone parameters determined to have statistically significant differences were highest in the tertile with the highest plasma level of n-3 PUFAs.

Table 3. Bone parameters by tertile of plasma *n*-3 PUFA (C18:3 (n3) Linolenic acid (ALA) + C20:5 (n3) Eicosapentenoic acid (EPA) + C22:6 (n3) Docosahexenoic acid (DHA)).

| | | Plasma Total *n*-3 PUFA Tertiles | | | | |
	Total Sample (*n* = 301) Median (IQR); *n* (%)	Lowest (<1.91) (*n* = 98) Median (IQR); *n* (%)	Middle (1.91–2.79) (*n* = 101) Median (IQR); *n* (%)	Highest (>2.79) (*n* = 102) Median (IQR); *n* (%)	*p* Value	Adjusted *p* Value *
Quantitative Bone Ultrasound						
BUA, dB/MHz	106 (14)	99 (10)	106 (15)	108 (14)	0.203	
SOS, m/s	1541 (35)	1526 (29)	1544 (38)	1541 (36)	0.531	
Volumetric BMD (mg/cm^3)						
Total density (mg/cm^3)	305.7 (73.7)	261.6 (71.5)	303 (81.4)	313.3 (80.3)	0.124	
Trabecular density (mg/cm^3)	163.1 (55.4)	131.7 (54.7)	164.5 (51.6)	161.5 (62.9)	0.72	
Cortical density (mg/cm^3)	418 (109.9)	350.3 (108.5) (a)	416 (123.7)	430.2 (98.9)	0.048	0.013
Bone morphometry (mm^2)						
Total area (mm^2)	300.4 (54)	272.2 (59.5)	300.4 (55.7)	301.8 (48.8)	0.373	
Trabecular area (mm^2)	135.1 (24.4)	122.2 (26.8)	135.1 (25.1)	136.6 (24)	0.273	
Cortical area (mm^2)	165.3 (29.9)	150 (32.8)	165.3 (30.6)	167.1 (29.2)	0.274	
Bone Mineral Density						
BMD L2-L4 (g/cm^2)	0.912 (0.205)	0.815 (0.186) (a)	0.907 (0.216)	0.936 (0.232)	0.011	0.043
BMD L2 (g/cm^2)	0.901 (0.204)	0.812 (0.171) (a)	0.902 (0.193)	0.933 (0.221)	0.023	0.123
BMD L3 (g/cm^2)	0.925 (0.229)	0.821 (0.197) (a)	0.924 (0.243)	0.949 (0.226)	0.033	0.074
BMD L4 (g/cm^2)	0.914 (0.204)	0.801 (0.184) (a)	0.895 (0.203)	0.954 (0.234)	0.004	0.021
Z-score (lumbar spine)	0.3 (1.6)	−0.6 (1.3) (a)	0.2 (1.7) (b)	0.6 (1.6)	0.001	0.034
T-score (lumbar spine)	−1.3 (2)	−2.3 (1.8) (a)	−1.3 (2)	−1.1 (2.2)	0.01	0.050
BMD Femoral trochanter (g/cm^2)	0.759 (0.13)	0.704 (0.107)	0.753 (0.185)	0.786 (0.149)	0.239	
BMD Femoral neck (g/cm^2)	0.611 (0.139)	0.529 (0.125) (a)	0.612 (0.136)	0.644 (0.138)	0.042	0.054
Z-score (hip)	0.5 (1.4)	−0.2 (1.1) (a)	0.4 (1.7)	0.8 (1.7)	0.035	0.064
T-score (hip)	−0.8 (1.3)	−1.3 (1.1)	−0.8 (1.7)	−0.5 (1.5)	0.234	
Bone health						
Normal	*n* = 103 (34.2%)	*n* = 24 (23.3%)	*n* = 38 (36.9%)	*n* = 41 (39.8%)	0.113	
Osteopenia	*n* = 145 (48.2%)	*n* = 55 (37.9%)	*n* = 43 (29.7%)	*n* = 47 (32.4%)		
Osteoporosis	*n* = 53 (17.6%)	*n* = 19 (35.8%)	*n* = 20 (37.7%)	*n* = 14 (26.4%)		

Between-group comparisons were performed using the Kruskal Wallis test or the Chi-square test as appropriate. (a) Poshoc analysis by Dunn's test, $p < 0.05$ vs. higher tertile. (b) Poshoc analysis by Dunn's test, $p < 0.05$ vs. higher tertile. * Adjusted by, age, years since menopause, BMI and vitamin D, calcium and energy intake (Quade's test).

A similar analysis was performed on plasma *n*-6 PUFA levels. Participants were classified into tertiles according to their plasma *n*-6 PUFA concentration; the results are shown in Table 4. Statistically significant differences were observed in the SOS as determined by QUS ($p = 0.045$), with the highest tertile of plasma *n*-6 PUFAs being significantly higher than the middle tertile. This observed difference did not remain statistically significant after adjustment for potential confounding factors.

The ratio of *n*-6 to *n*-3 PUFAs was also studied, in addition to its association with the bone parameters analyzed (Table 5). After participants were classified according to the tertile of the *n*-6/*n*-3 PUFA ratio, statistically significant differences in lumbar Z-score were observed ($p = 0.016$), but these differences did not remain after adjustment for potential confounding factors ($p = 0.166$).

The role of the studied SFAs is shown in Table 6. After the participants were classified according to their plasma SFA tertiles, no statistically significant differences between groups were observed in any of the groups studied.

Table 4. Bone parameters by tertile of plasma *n*-6 PUFA (C18:2 (n6) Linoleic acid (LA) + C20:4 (n6) Arachidonic acid (AA)).

	Total Sample (*n* = 301) Median (IQR); *n* (%)	**Plasma Total *n*-6 PUFA Tertiles** Lowest (<17.59) (*n* = 98) Median (IQR); *n* (%)	Middle (17.59–20.49) (*n* = 101) Median (IQR); *n* (%)	Highest (>20.49) (*n* = 102) Median (IQR); *n* (%)	*p* Value	Adjusted *p* Value *
Quantitative Bone Ultrasound						
BUA (dB/MHz)	106 (14)	105 (14)	106 (14)	107 (13)	0.599	
SOS (m/s)	1541 (35)	1540 (29)	1539 (36) (a)	1546 (35)	0.045	0.320
Volumetric BMD (mg/cm^3)						
Total density (mg/cm^3)	305.7 (73.7)	298.2 (82.4)	309.9 (74.7)	308 (65.9)	0.546	
Trabecular density (mg/cm^3)	163.1 (55.4)	162.1 (55.3)	164.6 (62.3)	164 (51.7)	0.664	
Cortical density (mg/cm^3)	418 (109.9)	409 (120.6)	429.3 (102.5)	423.1 (107.2)	0.422	
Bone morphometry (mm^2)						
Total area (mm^2)	300.4 (54)	299.5 (48.1)	304 (51.2)	294 (55)	0.123	
Trabecular area (mm^2)	135.1 (24.4)	135.3 (21)	136.8 (22.9)	132.1 (24.8)	0.11	
Cortical area (mm^2)	165.3 (29.9)	165.5 (26.1)	167 (28.2)	161.9 (30.3)	0.116	
Bone Mineral Density						
BMD L2-L4 (g/cm^2)	0.912 (0.205)	0.913 (0.252)	0.92 (0.187)	0.903 (0.194)	0.828	
BMD L2 (g/cm^2)	0.901 (0.204)	0.886 (0.221)	0.915 (0.196)	0.884 (0.185)	0.746	
BMD L3 (g/cm^2)	0.925 (0.229)	0.906 (0.266)	0.942 (0.197)	0.915 (0.197)	0.759	
BMD L4 (g/cm^2)	0.914 (0.204)	0.926 (0.258)	0.913 (0.186)	0.916 (0.178)	0.873	
Z-score (lumbar spine)	0.3 (1.6)	0.3 (2.2)	0.3 (1.3)	0.1 (1.5)	0.694	
T-score (lumbar spine)	−1.3 (2)	−1.3 (2.4)	−1.3 (1.8)	−1.4 (1.8)	0.856	
BMD Femoral trochanter (g/cm^2)	0.759 (0.13)	0.756 (0.135)	0.756 (0.121)	0.762 (0.149)	0.855	
BMD Femoral neck (g/cm^2)	0.611 (0.139)	0.607 (0.148)	0.622 (0.12)	0.606 (0.135)	0.645	
Z-score (hip)	0.5 (1.4)	0.6 (1.3)	0.5 (1.5)	0.5 (1.4)	0.95	
T-score (hip)	−0.8 (1.3)	−0.8 (1.2)	−0.8 (1.3)	−0.7 (1.4)	0.856	
Bone health						
Normal	*n* = 103 (34.2%)	*n* = 38 (36.9%)	*n* = 32 (31.1%)	*n* = 33 (32%)	0.390	
Osteopenia	*n* = 145 (48.2%)	*n* = 40 (27.6%)	*n* = 50 (34.5%)	*n* = 55 (37.9%)		
Osteoporosis	*n* = 53 (17.6%)	*n* = 20 (37.7%)	*n* = 19 (35.8%)	*n* = 14 (26.4%)		

Between-group comparisons were performed using the Kruskal Wallis test or the Chi-square test as appropriate. (a) Poshoc analysis by Dunn's test, *p* < 0.05 vs. higher tertile. * Adjusted by, age, years since menopause, BMI and vitamin D, calcium and energy intake (Quade's test).

Table 7 shows the results for the studied MUFAs. After the participants were classified on the basis of tertiles, statistically significant differences were observed in total (*p* = 0.02), trabecular (*p* = 0.014) and cortical (*p* = 0.014) bone morphometry. These statistically significant differences indicated a smaller area (mm^2) in those women belonging to the highest tertile of plasma MUFAs and remained statistically significant after adjustment for potential confounding factors (Table 7). In the group of women with high plasma MUFAs, the percentage of osteopenic women was also significantly increased.

When all measured PUFAs were pooled and participants were classified based on their plasma tertile (Table 8), statistically significant differences were observed in the SOS as determined by QUS (*p* = 0.004), but these differences did not remain statistically significant after adjustment for potential confounding factors.

Table 5. Bone parameters by tertile of plasma *n*-6 PUFA/*n*-3 PUFA ratio.

	Total Sample (*n* = 301) Median (IQR); *n* (%)	Plasma Total *n*-6/*n*-3 PUFA Tertiles			*p* Value	Adjusted *p* Value *
		Lowest (<6.79) (*n* = 98) Median (IQR); *n* (%)	Middle (6.79–10.27) (*n* = 101) Median (IQR); *n* (%)	Highest (>10.27) (*n* = 102) Median (IQR); *n* (%)		
Quantitative Bone Ultrasound						
BUA (dB/MHz)	106 (14)	105 (14)	106 (14)	107 (13)	0.683	
SOS (m/s)	1541 (35)	1540 (29)	1539 (36)	1546 (35)	0.982	
Volumetric BMD (mg/cm^3)						
Total density (mg/cm^3)	305.7 (73.7)	298.2 (82.4)	309.9 (74.7)	308 (65.9)	0.154	
Trabecular density (mg/cm^3)	163.1 (55.4)	162.1 (55.3)	164.6 (62.3)	164 (51.7)	0.875	
Cortical density (mg/cm^3)	418 (109.9)	409 (120.6)	429.3 (102.5)	423.1 (107.2)	0.099	
Bone morphometry (mm^2)						
Total area (mm^2)	300.4 (54)	299.5 (48.1)	304 (51.2)	294 (55)	0.966	
Trabecular area (mm^2)	135.1 (24.4)	135.3 (21)	136.8 (22.9)	132.1 (24.8)	0.88	
Cortical area (mm^2)	165.3 (29.9)	165.5 (26.1)	167 (28.2)	161.9 (30.3)	0.883	
Bone Mineral Density						
BMD L2-L4 (g/cm^2)	0.912 (0.205)	0.913 (0.252)	0.92 (0.187)	0.903 (0.194)	0.152	
BMD L2 (g/cm^2)	0.901 (0.204)	0.886 (0.221)	0.915 (0.196)	0.884 (0.185)	0.247	
BMD L3 (g/cm^2)	0.925 (0.229)	0.906 (0.266)	0.942 (0.197)	0.915 (0.197)	0.3	
BMD L4 (g/cm^2)	0.914 (0.204)	0.926 (0.258)	0.913 (0.186)	0.916 (0.178)	0.05	
Z-score (lumbar spine)	0.3 (1.6)	0.3 (2.2) (a)	0.3 (1.3) (b)	0.1 (1.5)	0.016	0.166
T-score (lumbar spine)	−1.3 (2)	−1.3 (2.4)	−1.3 (1.8)	−1.4 (1.8)	0.132	
BMD Femoral trochanter (g/cm^2)	0.759 (0.13)	0.756 (0.135)	0.756 (0.121)	0.762 (0.149)	0.48	
BMD Femoral neck (g/cm^2)	0.611 (0.139)	0.607 (0.148)	0.622 (0.12)	0.606 (0.135)	0.157	
Z-score (hip)	0.5 (1.4)	0.6 (1.3)	0.5 (1.5)	0.5 (1.4)	0.081	
T-score (hip)	−0.8 (1.3)	−0.8 (1.2)	−0.8 (1.3)	−0.7 (1.4)	0.413	
Bone health						
Normal	*n* = 103 (34.2%)	*n* = 38 (36.9%)	*n* = 39 (37.9%)	*n* = 26 (25.2%)	0.194	
Osteopenia	*n* = 145 (48.2%)	*n* = 42 (29%)	*n* = 46 (31.7%)	*n* = 57 (39.3%)		
Osteoporosis	*n* = 53 (17.6%)	*n* = 19 (35.8%)	*n* = 15 (28.3%)	*n* = 19 (25.8%)		

Between-group comparisons were performed using the Kruskal Wallis test or the Chi-square test as appropriate. (a) Poshoc analysis by Dunn's test, *p* < 0.05 vs. higher tertile. (b) Poshoc analysis by Dunn's test, *p* < 0.05 vs. higher tertile. * Adjusted by, age, years since menopause, BMI and vitamin D, calcium and energy intakes (Quade's test).

Table 6. Bone parameters by tertile of plasma Saturated Fatty Acids (SFA).

	Total Sample (*n* = 301) Median (IQR); *n* (%)	Plasma Total SFA Tertiles			*p* Value
		Lowest (<65.35) (*n* = 99) Median (IQR); *n* (%)	Middle (65.35–69.28) (*n* = 100) Median (IQR); *n* (%)	Highest (>69.28) (*n* = 102) Median (IQR); *n* (%)	
Quantitative Bone Ultrasound					
BUA (dB/MHz)	106 (14)	107 (12)	106 (13)	105 (16)	0.604
SOS (m/s)	1541 (35)	1549 (35)	1540 (31)	1540 (32)	0.052
Volumetric BMD (mg/cm^3)					
Total density (mg/cm^3)	305.7 (73.7)	302.4 (68.4)	303.2 (71.5)	308 (80.7)	0.726
Trabecular density (mg/cm^3)	163.1 (55.4)	163.4 (50)	162.4 (60.8)	164.3 (60.6)	0.902
Cortical density (mg/cm^3)	418 (109.9)	414 (109.7)	419 (109.4)	426.5 (117.4)	0.653
Bone morphometry (mm^2)					
Total area (mm^2)	300.4 (54)	296.9 (48.4)	303.1 (53.6)	302 (53.2)	0.154
Trabecular area (mm^2)	135.1 (24.4)	133 (21.6)	136.4 (24.2)	136 (22.2)	0.135
Cortical area (mm^2)	165.3 (29.9)	163.6 (26.8)	166.7 (29.4)	166.6 (27.5)	0.142
Bone Mineral Density					
BMD L2-L4 (g/cm^2)	0.912 (0.205)	0.901 (0.18)	0.917 (0.204)	0.926 (0.225)	0.738
BMD L2 (g/cm^2)	0.901 (0.204)	0.88 (0.186)	0.894 (0.188)	0.922 (0.213)	0.862
BMD L3 (g/cm^2)	0.925 (0.229)	0.913 (0.204)	0.925 (0.225)	0.946 (0.241)	0.776
BMD L4 (g/cm^2)	0.914 (0.204)	0.896 (0.171)	0.919 (0.223)	0.924 (0.225)	0.709
Z-score (lumbar spine)	0.3 (1.6)	0.1 (1.4)	0.3 (1.5)	0.4 (1.9)	0.697
T-score (lumbar spine)	−1.3 (2)	−1.4 (1.7)	−1.3 (1.9)	−1.2 (2.2)	0.727
BMD Femoral trochanter (g/cm^2)	0.759 (0.13)	0.756 (0.157)	0.757 (0.121)	0.771 (0.149)	0.715
BMD Femoral neck (g/cm^2)	0.611 (0.139)	0.607 (0.135)	0.609 (0.137)	0.625 (0.145)	0.486
Z-score (hip)	0.5 (1.4)	0.5 (1.6)	0.5 (1.2)	0.5 (1.5)	0.751
T-score (hip)	−0.8 (1.3)	−0.8 (1.4)	−0.7 (1.1)	−0.7 (1.6)	0.664

Table 6. *Cont.*

	Total Sample (n = 301) Median (IQR); n (%)	Plasma Total SFA Tertiles			p Value
		Lowest (<65.35) (n = 99) Median (IQR); n (%)	Middle (65.35–69.28) (n = 100) Median (IQR); n (%)	Highest (>69.28) (n = 102) Median (IQR); n (%)	
Bone health					
Normal	n = 103 (34.2%)	n = 29 (28.2%)	n = 33 (32%)	n = 41 (39.8%)	0.100
Osteopenia	n = 145 (48.2%)	n = 57 (39.3%)	n = 49 (33.8%)	n = 39 (26.9%)	
Osteoporosis	n = 53 (17.6%)	n = 13 (24.5%)	n = 18 (34%)	n = 22 (41.5%)	

Between-group comparisons were performed using the Kruskal Wallis test or the Chi-square test as appropriate. SFA (Saturated Fatty Acid: C8:0 Caprylic acid, + C10:0 Decanoic acid, + C12:0 Lauric acid, + C14:0 Myristic acid, + C16:0 Palmitic acid, + C18:0 Stearic acid, + C20:0 Arachidic acid, + C22:0 Docosanoic acid, + C24:0 Lignoceric acid).

Table 7. Bone parameters by tertile of plasma Monounsaturated Fatty Acids (MUFA) (C16:1 Palmitoleic acid + C18:1 cis (n9) Oleic acid + C22:1 (n9) Erucic acid).

	Total Sample (n = 301) Median (IQR); n (%)	Plasma Total MUFA Tertiles			p Value	Adjusted p Value *
		Lowest (<9.66) (n = 99) Median (IQR); n (%)	Middle (9.66–12.37) (n = 100) Median (IQR); n (%)	Highest (>12.37) (n = 102) Median (IQR); n (%)		
Quantitative Bone Ultrasound						
BUA (dB/MHz)	106 (14)	107 (17)	105 (12)	106 (11)	0.395	
SOS (m/s)	1541 (35)	1540 (33)	1541 (38)	1544 (35)	0.836	
Volumetric BMD (mg/cm^3)						
Total density (mg/cm^3)	305.7 (73.7)	309.4 (79.9)	306.7 (83)	294.7 (69.1)	0.409	
Trabecular density (mg/cm^3)	163.1 (55.4)	159.1 (59.4)	164.6 (57.8)	164.5 (53.6)	0.885	
Cortical density (mg/cm^3)	418 (109.9)	434.1 (109.2)	420.6 (121.5)	408.5 (106.5)	0.141	
Bone morphometry (mm^2)						
Total area (mm^2)	300.4 (54)	307 (48.8) (a)	302.2 (63.9) (b)	290.7 (41.8)	0.02	0.012
Trabecular area (mm^2)	135.1 (24.4)	137.8 (21.2) (a)	135.8 (28.5) (b)	130.9 (18.8)	0.014	0.006
Cortical area (mm^2)	165.3 (29.9)	169.2 (25.8) (a)	166.4 (35.4) (b)	160.3 (23)	0.014	0.006
Bone Mineral Density						
BMD L2-L4 (g/cm^2)	0.912 (0.205)	0.927 (0.216)	0.934 (0.228)	0.883 (0.173)	0.078	
BMD L2 (g/cm^2)	0.901 (0.204)	0.919 (0.22)	0.912 (0.226)	0.865 (0.161)	0.07	
BMD L3 (g/cm^2)	0.925 (0.229)	0.941 (0.214)	0.953 (0.245)	0.896 (0.162)	0.056	
BMD L4 (g/cm^2)	0.914 (0.204)	0.933 (0.205)	0.923 (0.231)	0.879 (0.179)	0.26	
Z-score (lumbar spine)	0.3 (1.6)	0.4 (1.9)	0.3 (1.6)	0 (1.4)	0.308	
T-score (lumbar spine)	−1.3 (2)	−1.2 (2)	−1.1 (2.2)	−1.6 (1.7)	0.075	
BMD Femoral trochanter (g/cm^2)	0.759 (0.13)	0.779 (0.171)	0.759 (0.127)	0.744 (0.139)	0.219	
BMD Femoral neck (g/cm^2)	0.611 (0.139)	0.63 (0.135)	0.594 (0.143)	0.609 (0.138)	0.229	
Z-score (hip)	0.5 (1.4)	0.6 (1.7)	0.5 (1.3)	0.4 (1.4)	0.586	
T-score (hip)	−0.8 (1.3)	−0.6 (1.7)	−0.7 (1.2)	−0.9 (1.3)	0.179	
Bone health						
Normal	n = 103 (34.2%)	n = 37 (35.9%)	n = 42 (40.8%)	n = 24 (23.3%)	0.020	
Osteopenia	n = 145 (48.2%)	n = 41 (28.3%)	n = 42 (29%)	n = 62 (42.8%)		
Osteoporosis	n = 53 (17.6%)	n = 21 (39.6%)	n = 16 (30.2%)	n = 16 (30.2%)		

Between-group comparisons were performed using the Kruskal Wallis test or the Chi-square test as appropriate. (a) Poshoc analysis by Dunn's test, $p < 0.05$ vs. higher tertile. (b) Poshoc analysis by Dunn's test, $p < 0.05$ vs. higher tertile. * Adjusted by, age, years since menopause, BMI and vitamin D, calcium and energy intakes (Quade's test).

Table 8. Bone parameters by tertile of total plasma PUFA.

	Total Sample (n = 301) Median (IQR); n (%)	Lowest (<20.28) (n = 99) Median (IQR); n (%)	Middle (20.28–22.95) (n = 101) Median (IQR); n (%)	Highest (>22.95) (n = 101) Median (IQR); n (%)	p Value	Adjusted p Value *
Quantitative Bone Ultrasound						
BUA (dB/MHz)	106 (14)	105 (15)	106 (15)	107 (14)	0.125	
SOS (m/s)	1541 (35)	1538 (28) (a)	1541 (31) (b)	1551 (38)	0.004	0.588
Volumetric BMD (mg/cm^3)						
Total density (mg/cm^3)	305.7 (73.7)	303.2 (82.3)	307.3 (80.3)	306.3 (65.70)	0.671	
Trabecular density (mg/cm^3)	163.1 (55.4)	162.2 (59.6)	166.5 (57.1)	161.4 (51.5)	0.49	
Cortical density (mg/cm^3)	418 (109.9)	414.2 (124.8)	429.3 (103.5)	421.5 (108.8)	0.517	
Bone morphometry (mm^2)						
Total area (mm^2)	300.4 (54)	296.6 (47.4)	302.8 (53.3)	298 (57.8)	0.512	
Trabecular area (mm^2)	135.1 (24.4)	133.3 (21.2)	136.8 (24)	134 (26.1)	0.405	
Cortical area (mm^2)	165.3 (29.9)	163.3 (26.3)	167 (28.9)	164.7 (31.7)	0.414	
Bone Mineral Density						
BMD L2-L4 (g/cm^2)	0.912 (0.205)	0.883 (0.215)	0.934 (0.185)	0.904 (0.198)	0.14	
BMD L2 (g/cm^2)	0.901 (0.204)	0.873 (0.209)	0.923 (0.211)	0.901 (0.185)	0.216	
BMD L3 (g/cm^2)	0.925 (0.229)	0.898 (0.251)	0.951 (0.206)	0.916 (0.21)	0.144	
BMD L4 (g/cm^2)	0.914 (0.204)	0.877 (0.244)	0.939 (0.182)	0.914 (0.181)	0.094	
Z-score (lumbar spine)	0.3 (1.6)	0.1 (2)	0.4 (1.2)	0.2 (1.4)	0.305	
T-score (lumbar spine)	−1.3 (2)	−1.6 (2)	−1.1 (1.8)	−1.4 (1.9)	0.173	
BMD Femoral trochanter (g/cm^2)	0.759 (0.13)	0.745 (0.13)	0.78 (0.103)	0.759 (0.177)	0.178	
BMD Femoral neck (g/cm^2)	0.611 (0.139)	0.602 (0.137)	0.625 (0.118)	0.604 (0.141)	0.291	
Z-score (hip)	0.5 (1.4)	0.4 (1.4)	0.6 (1.4)	0.5 (1.6)	0.313	
T-score (hip)	−0.8 (1.3)	−0.9 (1.3)	−0.6 (1.1)	−0.8 (1.6)	0.197	
Bone health						
Normal	n = 103 (34.2%)	n = 31 (30.1)	n = 39 (37.9%)	n = 33 (32.0%)	0.428	
Osteopenia	n = 145 (48.2%)	n = 46 (31.7%)	n = 45 (31%)	n = 54 (37.2%)		
Osteoporosis	n = 53 (17.6%)	n = 22 (41.5%)	n = 17 (32.1%)	n = 14 (26.4%)		

Between-group comparisons were performed using the Kruskal Wallis test or the Chi-square test as appropriate. (a) Poshoc analysis by Dunn's test, $p < 0.05$ vs. higher tertile. (b) Poshoc analysis by Dunn's test, $p < 0.05$ vs. higher tertile. * Adjusted by, age, years since menopause, BMI and vitamin D, calcium and energy intakes (Quade's test). Total plasma PUFA: C18:3 (n3) Linolenic acid (ALA) + C20:5 (n3) Eicosapentenoic acid (EPA) + C22:6 (n3) Docosahexenoic acid (DHA) + C18:2 (n6) Linoleic acid (LA) + C20:4 (n6) Arachidonic acid (AA).

3.3. Correlation Study

To further investigate the associations between BMD in the women studied and the fatty acids studied, we conducted an analysis using bivariate correlations and then partial correlations adjusted for age; years since menopause; BMI; and intake levels of vitamin D, calcium, and energy. The results are shown in Table 9. Statistically significant positive correlations were observed between BMD and plasma *n*-3 PUFA levels at the lumbar spine (r = 0.157; $p = 0.006$), L2 (r = 0.143; $p = 0.013$), L3 (r = 0.128; $p = 0.026$ and L4 (r = 0.178; $p = 0.002$) and femoral neck (r = 0.153; $p = 0.008$). Statistically significant and negative bivariate correlations were also reported between BMD and plasma *n*-6/*n*-3 PUFA levels at the lumbar spine (r = −0.121; $p = 0.035$), L4 (r = −0.156; $p = 0.007$) and femoral neck (r = −0.124; $p = 0.031$). Subsequent analysis of the partial correlations adjusted for the confounding factors studied showed that those observed for lumbar spine BMD and plasma levels of *n*-3 PUFAs remained statistically significant and positive (total spine (r = 0.150; $p = 0.014$), L2 (r = 0.141; $p = 0.021$), L3 (r = 0.129; $p = 0.035$), L4 (r = 0.170; $p = 0.005$) and femoral neck (r = 0.143; $p = 0.019$)). The bivariate correlations observed between plasma *n*-6/*n*-3 PUFA levels and the lumbar spine and femoral neck did not remain statistically significant ($p = 0.086$ and $p = 0.073$, respectively), although the correlation with L4 BMD did remain significant (r = −0.139; $p = 0.023$).

Table 9. Bivariate and partial correlations between plasma fatty acids and bone mineral density at either the lumbar spine or the hips.

Variables		BMD L2-L4 (g/cm^2)	BMD L2 (g/cm^2)	BMD L3 (g/cm^2)	BMD L4 (g/cm^2)	BMD Femoral Trochanter (g/cm^2)	BMD Femoral Neck (g/cm^2)
Plasma total *n*-3 PUFA	Spearman Rho	0.157	0.143	0.128	0.178	0.11	0.153
	p Value	0.006	0.013	0.026	0.002	0.056	0.008
	Adjusted coefficient *	0.150	0.141	0.129	0.170		0.143
	p Value	0.014	0.021	0.035	0.005		0.019
Plasma total *n*-6 PUFA	Spearman Rho	0.015	0.041	0.026	−0.013	0.027	−0.021
	p Value	0.802	0.481	0.652	0.819	0.636	0.721
Plasma *n*-6/*n*-3 ratio	Spearman Rho	−0.121	−0.094	−0.094	−0.156	−0.069	−0.127
	p Value	0.035	0.102	0.102	0.007	0.23	0.035
	Adjusted coefficient *	−0.105			−0.139		−0.110
	p Value	0.086			0.023		0.073
Plasma total SFA	Spearman Rho	0.031	0.016	0.026	0.034	0.027	0.057
	p Value	0.59	0.776	0.659	0.552	0.639	0.324
Plasma total MUFA	Spearman Rho	−0.106	−0.113	−0.097	−0.082	−0.101	−0.091
	p Value	0.066	0.05	0.092	0.155	0.08	0.114
Plasma total PUFA	Spearman Rho	0.056	0.077	0.058	0.037	0.056	0.012
	p Value	0.336	0.18	0.312	0.517	0.33	0.829

* Partial non parametric correlations adjusted by, age, years since menopause, BMI and vitamin D, calcium and energy intakes.

Partial correlations between the fatty acid profile and BMD at the spine or hip level are shown in Supplementary Table S1. After adjustment for potential confounders, the correlations observed between ALA and BMD L4 (r = 0. 123; *p* = 0.044) and between DHA and BMD at the lumbar spine (r = 0.135; *p* = 0.026), L2 (r = 0.148; *p* = 0.015), L3 (r = 0.122; *p* = 0.045), L4 (r = 0.136; *p* = 0.0.26), and femoral neck (r = 0.132; *p* = 0.031) remained statistically significant. A statistically significant partial correlation was also observed between erucic acid and BMD at the femoral neck level (r = −0.122; *p* = 0.035).

3.4. Multiple Linear Regression Study: Identification of Predictors

Independent relationships between BMD and the fatty acids studied were also explored using multiple linear regression including age (years), BMI (kg/m^2), vitamin D (mcg/day), calcium intake (mg/day), energy (kcal/day), plasma total *n*-3 PUFA, plasma total *n*-6 PUFA, plasma *n*-6/*n*-3 ratio, plasma total SFA, plasma total MUFA and plasma total PUFA as explanatory variables. An independent statistically significant positive relationship was observed between BMD at the spine and BMI (β = 0.288; *p* = 0.001) as well as plasma total *n*-3 PUFAs (β = 0.155; *p* = 0.009), while a negative association was observed with age (β = −0.163; *p* = 0.006). The proposed model accounted for up to 11% of the variability associated with spine BMD in the study group (Table 10). Subsequently, we proceeded to repeat the analysis after classifying the participants based on the WHO osteopenia/osteoporosis criteria (Table 10). In the group of women with normal T-scores, only *n*-3 PUFA plasma levels (β = 0.273; *p* = 0.001) functioned as a positive predictor of BMD at the lumbar spine level, yielding a model that explained only 7% of the observed variability. In osteopenic women, the predictor included in the model was daily energy intake (β = 0.226; *p* = 0.011). Finally, in osteoporotic women, the model included a positive relationship with plasma *n*-3 PUFA levels (β = 0.290; *p* = 0.024) and a negative relationship with plasma MUFA levels (β = −0.411; *p* = 0.002). This model explained 28% of the variability observed in BMD at the lumbar level.

A parallel analysis was performed for BMD at the femoral neck level (Table 11). Overall, in the study sample, BMI β = 0.394; *p* ≤ 0.001) and plasma *n*-3 PUFA levels (β = 0.146; *p* = 0.009) were positive predictors of BMD at the femoral neck level. The global model accounted for a total of 18% of the reported variability (Table 11). After the subjects were categorized according to the WHO osteopenia/osteoporosis criteria, no predictors of BMD at the femoral neck level were identified in the group of osteoporotic women (Table 11). In the group of women with normal T-scores, the positive predictors observed were BMI (β = 0.320; p = 0.002) and plasma *n*-3 PUFAs (β =0.245; p = 0.015). This generated a model explaining 18% of the observed variability. In the group of osteoporotic women, only BMI β = 0.299; p = 0.001) was incorporated into the model as a positive predictor of BMD at the femoral neck level. The model was able to explain only 9% of the observed variability.

Table 10. Multiple linear regression analysis for the BMD at the lumbar spine.

Total sample. Spine BMD				
Optimal model	R2	Adjusted R2	F	*p*
	0.126	0.116	12.552	<0.001
Selected independent variables		standardized B	t	*p*
BMI (kg/m^2)		0.288	4.961	<0.001
Age (years)		−0.163	−2.769	0.006
Plasma total *n*-3 PUFA		0.155	2.643	0.009
Normal women. Spine BMD				
Optimal model	R2	Adjusted R2	F	*p*
	0.074	0.064	6.922	0.01
Selected independent variables		standardized B	t	*p*
Plasma total *n*-3 PUFA		0.273	2.631	0.01
Osteopenic women. Spine BMD				
Optimal model	R2	Adjusted R2	F	*p*
	0.051	0.044	6.737	0.011
Selected independent variables		standardized B	t	*p*
Energy (kcal/day)		0.226	2.596	0.011
Osteoporotic women. Spine BMD				
Optimal model	R2	Adjusted R2	F	*p*
	0.283	0.253	9.282	<0.001
Selected independent variables		standardized B	t	*p*
Plasma total MUFA		−0.411	−3.303	0.002
Plasma total *n*-3 PUFA		0.290	2.323	0.024

Predictors: Age (years), Years since menopause (years), BMI (kg/m^2), Vitamin D (µg/day), Calcium intake (mg/day), Energy (kcal/day), Plasma total *n*-3 PUFA, Plasma total *n*-6 PUFA, Plasma *n*-6/*n*-3 ratio, Plasma total SFA, Plasma total MUFA and Plasma total PUFA.

Table 11. Multiple linear regression analysis for the BMD at the femoral neck.

Total sample. Femoral neck BMD				
Optimal model	R2	Adjusted R2	F	*p*
	0.185	0.179	29.946	<0.001
Selected independent variables		standardized B	t	*p*
BMI (kg/m^2)		0.394	7.068	<0.001
Plasma total *n*-3 PUFA		0.146	2.615	0.009
Normal women sample. Femoral neck BMD				
Optimal model	R2	Adjusted R2	F	*p*
	0.187	0.168	9.797	<0.001
Selected independent variables		standardized B	t	*p*
BMI (kg/m^2)		0.320	3.227	0.002
Plasma total *n*-3 PUFA		0.245	2.477	0.015
Osteopenic women sample. Femoral neck BMD				
Optimal model	R2	Adjusted R2	F	*p*
	0.090	0.082	12.410	0.001
Selected independent variables		standardized B	t	*p*
BMI (kg/m^2)		0.299	3.523	0.001

Predictors: Age, years. Years since menopause, years. BMI (kg/m^2), Vitamin D (µg/day), Calcium intake (mg/day), Energy (kcal/day), Plasma total *n*-3 PUFA, Plasma total *n*-6 PUFA, Plasma *n*-6/*n*-3 ratio, Plasma total SFA, Plasma total MUFA and Plasma total PUFA.

3.5. Logistic Regression Analysis for Low Bone Mass (T-score ≤ 1)

Independent risk factors for low BMD (T-score ≤ 1) were determined by logistic regression analysis. The analysis identified a higher BMI (OR = 0.893; 95% CI 0.841–0.948), $p < 0.001$) and a higher level of plasma *n*-3 PUFAs (OR = 0.751; 95% CI 0.587–0.960, $p = 0.022$) as protective factors against low bone mass. No other statistically significant factors were observed in the study sample (Table 12). Regarding the fatty acid profile and the risk of low bone mass, the results are shown in Supplementary Table S2, and no statistically significant results were observed.

Table 12. Logistic regression of predictors associated with low BMD (T ≤ −1 score).

	Univariate			Multivariate		
	OR	**95% CI**	*p* **Value**	**OR**	**95% CI**	*p* **Value**
Age, years	1.040	0.997–1.084	0.068			
Years since menopause, years	0.987	0.956–1.020	0.439			
BMI (kg/m^2)	0.905	0.858–0.954	<0.001	0.893	0.841–0.948	$p < 0.001$
Vitamin D (µg/day)	1.000	0.993–1.006	0.895			
Calcium intake (mg/day)	1.000	1.000–1.001	0.785			
Energy (kcal/day)	1.000	1.000–1.000	0.676			
Plasma total *n*-3 PUFA	0.791	0.635–0.984	0.035	0.751	0.587–0.960	0.022
Plasma total *n*-6 PUFA	1.014	0.953–1.079	0.655			
Plasma *n*-6/*n*-3 ratio	1.024	0.998–1.050	0.069			
Plasma total SFA	0.982	0.942–1.023	0.379			
Plasma total MUFA	1.073	0.993–1.158	0.075			
Plasma total PUFA.	0.995	0.935–1.058	0.874			

OR: Odds ratio; CI: Confidence interval; Reference level_ T-score at either spine or femoral neck > −1. Multivariate model adjusted by: Age (years), Years since menopause (years) BMI (kg/m^2), Vitamin D (µg/day), Calcium intake (mg/day), Energy (kcal/day).

4. Discussion

To the best of our knowledge, this is the first investigation of the association between the plasma fatty acid profile and bone density as determined by QUS, pQCT, and DXA. No consistent associations have been established between the levels of most of the fatty acids studied by QUS and pQCT. However, we have identified a stable association between the highest level of plasma *n*-3 PUFAs (ALA + EPA + DHA) and BMD at the level of the lumbar spine, but the individual role that the plasma level of each of these fatty acids might play in relation to BMD is difficult to interpret since each of these omega-3 fatty acids has different functions. Combined analysis reported here lead us to believe that they confirm previous findings obtained from observational studies in which it was observed that dietary intake of *n*-3 PUFAs acids was associated with better bone health and even with a reduction in the risk of fractures. Furthermore, these results lead us to consider that most of the fatty acids studied do not seem to have a notable impact and that future research should focus on deepening the role of *n*-3 PUFAs. Our results indicate that there is a consistent association with BMD at the lumbar level and probably a weaker association with BMD at the femoral neck area, but nevertheless, their higher concentration is associated with a lower risk of developing low bone mass in our sample of postmenopausal Spanish women.

Very few studies have established a relationship between the plasma fatty acid profile and bone density; most of these studies are focused on the study of *n*-3 and *n*-6 PUFAs and BMD determined by DXA. However, more data are available on associations between dietary intake of these fatty acids and BMD as if diets are rich in ALA + EPA + DHA then it is expected a higher level of plasma *n*-3 PUFAs and particularly, it has been shown that fatty acids are incorporated into plasma phospholipids after intake [29]. Hence, appropriate nutritional management, such as intake of *n*-3 fatty acids, may be a strategy to maximise peak bone mass in the female hip [30]. The interrelations of total and individual dietary PUFA intake with bone health are intricate [5,31–33] and may be further affected by nutrient-nutrient interactions, as well as by the ratio of *n*-6 to *n*-3 fatty acid intake. Given that the major dietary sources of *n*-3 PUFAs are foods of marine origin, different studies have examined the associations between their consumption and BMD. In 2011, Farina et al. [19] observed that the protective effects on BMD at the femoral neck level associated with a high intake of AA could depend on the intake levels of EPA + DHA, concluding that fish consumption can protect against bone loss and that these protective effects could also depend on a fine relationship between the levels of different fatty acids. The findings regarding fish consumption and bone health were further supported by QUS in premenopausal Spanish women [34]. In a further continuation of the original work by Farina et al. in 2012 [5], these authors analyzed relationships between plasma levels of certain fatty acids and BMD as well as risk of fracture. That study is believed to be the first work to address the association between plasma fatty acids and BMD determined by DXA. Their results suggested protective effects of plasma AA on femoral neck BMD in men, as well as protective effects of plasma AA against hip fracture risk in both women and men. Additionally, their results showed possible detrimental effects of plasma LA on BMD or hip fracture risk while identifying BMI as a possible confounding factor to be taken into account in these analyses. Their study suggested a possible more subtle effect of plasma DHA on BMD in older women and men. Overall, these results initially reported by the Farina et al. group have been confirmed by ours, suggesting that plasma levels of *n*-3 PUFAs appear to be associated with increased BMD at both the lumbar and hip levels, and DHA appears to correlate positively with BMD after adjustment for BMI and other potential confounders in our sample of postmenopausal women. Additional studies focusing on the analysis of osteoporotic fracture risk have also confirmed that higher plasma PUFA concentrations in old age are associated with a lower risk of osteoporotic fracture in men, but these findings were not replicated in women, as they did not reach statistical significance [35]. In men, plasma *n*-3 fatty acids and specifically plasma EPA seem to be relevant among PUFAs, while *n*-6 fatty acids in women may be associated with an increased risk of fracture.

However, focusing again on the nutritional aspect, it is true that not all studies have observed positive associations between PUFA intake and bone health. A long-term follow-up (two years) of patients with osteoarthritis of the knee supplemented with omega-3 fish oil did not demonstrate any efficacy of omega-3 fatty acids on bone loss in n = 202 Australian men and women aged $\geq$ 40 [36]. The findings in postmenopausal women enrolled in the Women's Health Initiative are also remarkable [17]. Focusing on fracture risk, that study found that a higher dietary intake of SFAs was associated with a higher risk of hip fracture in postmenopausal women, while a higher intake of PUFAs and MUFAs was associated with a lower fracture risk. The authors reported no associations between total *n*-3 PUFA or ALA intake and fracture risk, but, strikingly, there was a slight increase in fracture risk associated with higher EPA + DHA intake. Regarding *n*-6 PUFAs, postmenopausal women with higher *n*-6 PUFA intake had a lower fracture risk. Although we did not determine the risk of fracture in our study, our results do seem to indicate that no negative effect of SFAs can be assumed, having detected only a small negative association between a higher plasma MUFA level and bone morphometry. These results, due to their preliminary nature, require additional studies to confirm that the association is not spurious. Neither have we observed negative or positive effects associated with a higher plasma level of *n*-6 PUFAs. Even the *n*-6/*n*-3 ratio, which is presumed to be more likely than individual classes of PUFAs to affect skeletal health [1,37,38], was not observed to have an effect in this study, although the Women's Health Initiative observed that a higher ratio (>6.43) (the ratio in our sample was 8.28 (5.85)) offered discrete protection against fractures.

Subsequent results obtained by measuring *n*-3 fatty acids in red blood cells in a sample from the Women's Health Initiative [39] suggested that more *n*-3 fatty acids in red blood cells (particularly ALA or EPA but not DHA) might be predictive of a lower risk of hip fracture. However, in this study, associations were not observed at baseline or after a 6-year follow-up. There was no relationship between red blood cell *n*-3 or *n*-6 fatty acids and BMD. The authors adjusted their analyses for potential confounders (alcohol consumption, total energy intake, total calcium intake, total vitamin D intake, and multivitamin use) and no changes were observed at longitudinal follow-up that would suggest an association between hip BMD and total *n*-3 or *n*-6 fatty acids in red blood cells. Similar results were observed when exploring the role of the *n*-6/*n*-3 ratio and the percentage change in BMD at longitudinal follow-up. By measuring *n*-3 FAs in red blood cells in Korean postmenopausal women [40], EPA, DHA, and total *n*-3 FAs were found to correlate with increased BMD of the femoral neck after adjusting for relevant confounders (age, years after menopause and height), whereas a higher *n*-6/*n*-3 FA ratio was correlated with decreased femoral neck BMD. Among young men, using an experimental approach similar to ours [3], it was also observed that concentrations of *n*-3 fatty acids, especially DHA, were positively associated with peak BMD in the whole body and spine and with bone accrual in the lumbar spine.

Regarding plasma fatty acids other than *n*-3 or *n*-6 PUFAs, in this study, statistically significant differences were observed according to the diagnosis of osteopenia/osteoporosis in C16:0 (palmitic acid), C24:0 (lignoceric acid) and C18:1 cis (n9) (oleic acid) after adjustment for potential confounding factors. For the first two, the level was significantly lower in women with osteopenia, while C18:1 cis (n9) (oleic acid) was more abundant in these women than in women with normal bone health according to the WHO osteopenia/osteoporosis criteria. There are previous results from a nutritional perspective indicating that the role of fatty acids in women's bone health could be limited to the situation of osteopenia, with their role becoming secondary once osteoporosis is established [20]. In any case, our results are contrary to those recently published in which palmitic acid levels were associated with increased odds of low BMD in Chinese adult women, but our findings will require in-depth analysis in future studies, given that other types of associations were not observed in our study and therefore could represent some type of spurious association [41].

Evidence from in vivo studies on the role of different dietary sources of *n*-3 PUFAs on skeletal development and bone quality has shown that dietary *n*-3 PUFAs contribute to improved mechanical and morphometric properties of bone, and bone quality [42] with possible protective mechanisms against bone loss during ageing, associated with inhibition of inflammation associated with bone resorption mediated by NF-κβ, p38MAPK or JNK leading to the regulation of pro-inflammatory cytokine release [43]. However, the current evidence from human studies is limited at best. Thus, *n*-3 PUFA fatty acids could maintain bone density in postmenopausal women, but the mechanism is not known. Different mechanisms have been proposed. Essential fatty acids, by regulating eicosanoids, leptin and IGF-1, are linked to the regulation of both growth and bone status [44,45]. Other possible underlying determinants have also been suggested as potential meditators of the relationship between plasma fatty acids and BMD. They involve the regulation of lipid mediators, inflammation, or oxidative stress [21,46–48], a critical determinant of the decrease in bone strength and mass [49]. As some fatty acids have been found to be related to the potential to modulate the levels of both inflammatory cytokines (such IL-6, IL-1beta, and TNF-alpha) and eicosanoids [50–52], it has been hypothesized that they might lower the generation of free radicals and oxidative stress and somehow diminish the age-related loss of bone mass [49]. Among the possible molecular mechanisms that could be regulated by certain fatty acids, one of the most active fields of research concerns mechanisms linked to the regulation of prostaglandin E2 biosynthesis [53,54], which modulates osteoclastogenesis to accelerate bone resorption through activation of the nuclear factor kappa-B pathway [55] and through regulation of the synthesis of osteoprotegerin [12,13,56].

A strength of our study is that it includes the measurement of plasma fatty acids, which provides an unbiased measure of exposure and is more accurate than the information obtained from dietary surveys. We recognize that the main limitations are the cross-sectional design of the study, which prevents us from establishing cause-effect relationships, and the use of convenience sampling, which could lead to biases derived from the composition of the sample and also prevents the generalization of our results. Furthermore, plasma fatty acids were measured at a single time point, whereas it would be more accurate to perform a follow-up over time to establish previous exposure; nevertheless, it appears that plasma *n*-3 PUFA levels remain fairly consistent over time [57]. Another important limitation is the lack of analysis of plasma levels of C22:5*n*-6 which is a clear indicator of omega 3 fatty acid deficiency [58]. Finally, as the entire population in this study was Caucasian, this demographic restriction may affect the generalizability of the results of our study.

5. Conclusions

In this single-center sample of postmenopausal Spanish women, we reported a significant positive and statistically independent association between BMD and plasma levels of *n*-3 PUFAs that highlights the physiological and biochemical relevance of plasma total omega 3 fatty acids. Longitudinal observational or randomized controlled studies are needed to further investigate any effect of *n*-3 PUFAs on bone health.

Supplementary Materials: The following are available online at https://www.mdpi.com/article/10.3390/nu13051454/s1: Table S1: Bivariate and partial correlations between plasma fatty acid profile and bone mineral density in the lumbar spine and the hips; Table S2: Logistic regression of fatty acids associated with low BMD (T-score ≤ -1).

Author Contributions: Conceptualization, R.R.-M., I.A. and J.M.M.; methodology, V.V., M.d.l.L.C.-M., P.R.-S., F.L.-E., A.S.-F., J.M.L.-G., L.M.P.-P. and J.D.P.-Z.; formal analysis, J.M.M., P.C.-C., R.R.-M. and I.A., writing—original draft preparation, J.M.M., P.C.-C., R.R.-M. and I.A.; supervision, J.M.M. All authors have read and agreed to the published version of the manuscript.

Funding: This research was funded by Junta de Extremadura, Consejería de Economía e Infraestructuras, Spain, Fondo Europeo de Desarrollo Regional, "Una manera de hacer Europa", grants number IB18042 and IB18044.

Institutional Review Board Statement: The study was conducted according to the guidelines of the Declaration of Helsinki, and approved by the Institutional Review Board of University of Extremadura (protocol code 84/2018 and date of approval 6 July 2018).

Informed Consent Statement: Informed consent was obtained from all subjects involved in the study.

Data Availability Statement: The dataset analyzed during the current study is not publicly available due to national data regulations and for ethical reasons, including that we do not have the explicit written consent of the study volunteers to make their deidentified data available at the end of the study. However, datasets and SPSS statistical analyses can be requested by sending a letter to the corresponding author.

Acknowledgments: The authors would like to thank Jose Luis Alonso Amado (RN) for his useful technical assistance. Technical and human support provided by Facility of Bioscience Applied Techniques of SAIUEx (financed by UEX, Junta de Extremadura, MICINN, FEDER and FSE) is acknowledged.

Conflicts of Interest: The authors declare no conflict of interest. The funders had no role in the design of the study; in the collection, analyses, or interpretation of data; in the writing of the manuscript, or in the decision to publish the results.

References

1. Weiss, L.A.; Barrett-Connor, E.; von Mühlen, D. Ratio of n–6 to n–3 Fatty Acids and Bone Mineral Density in Older Adults: The Rancho Bernardo Study. *Am. J. Clin. Nutr.* **2005**, *81*, 934–938. [CrossRef]
2. Eriksson, S.; Mellström, D.; Strandvik, B. Fatty Acid Pattern in Serum Is Associated with Bone Mineralisation in Healthy 8-Year-Old Children. *Br. J. Nutr.* **2009**, *102*, 407–412. [CrossRef] [PubMed]
3. Hogstrom, M.; Nordström, P.; Nordström, A. N-3 Fatty Acids Are Positively Associated with Peak Bone Mineral Density and Bone Accrual in Healthy Men: The NO$_2$ Study. *Am. J. Clin. Nutr.* **2007**, *85*, 803–807. [CrossRef] [PubMed]
4. Sánchez-Borrego, R.; von Schacky, C.; Osorio, M.J.A.; Llaneza, P.; Pinto, X.; Losa, F.; Navarro, M.C.; Lubián, D.; Mendoza, N. Recommendations of the Spanish Menopause Society on the Consumption of Omega-3 Polyunsaturated Fatty Acids by Postmenopausal Women. *Maturitas* **2017**, *103*, 71–77. [CrossRef] [PubMed]
5. Farina, E.K.; Kiel, D.P.; Roubenoff, R.; Schaefer, E.J.; Cupples, L.A.; Tucker, K.L. Plasma Phosphatidylcholine Concentrations of Polyunsaturated Fatty Acids Are Differentially Associated With Hip Bone Mineral Density and Hip Fracture in Older Adults: The Framingham Osteoporosis Study. *J. Bone Miner. Res.* **2012**, *27*, 1222–1230. [CrossRef]
6. Williams, C.M.; Burdge, G. Long-Chain *n*-3 PUFA: Plant v. Marine Sources. *Proc. Nutr. Soc.* **2006**, *65*, 42–50. [CrossRef]

7. Nawata, K.; Yamauchi, M.; Takaoka, S.; Yamaguchi, T.; Sugimoto, T. Association of N-3 Polyunsaturated Fatty Acid Intake with Bone Mineral Density in Postmenopausal Women. *Calcif. Tissue Int.* **2013**, *93*, 147–154. [CrossRef]
8. Järvinen, R.; Tuppurainen, M.; Erkkilä, A.T.; Penttinen, P.; Kärkkäinen, M.; Salovaara, K.; Jurvelin, J.S.; Kröger, H. Associations of Dietary Polyunsaturated Fatty Acids with Bone Mineral Density in Elderly Women. *Eur. J. Clin. Nutr.* **2012**, *66*, 496–503. [CrossRef]
9. Mangano, K.M.; Sahni, S.; Kerstetter, J.E.; Kenny, A.M.; Hannan, M.T. Polyunsaturated Fatty Acids and Their Relation with Bone and Muscle Health in Adults. *Curr. Osteoporos. Rep.* **2013**, *11*, 203–212. [CrossRef] [PubMed]
10. Haag, M.; Magada, O.N.; Claassen, N.; Böhmer, L.H.; Kruger, M.C. Omega-3 Fatty Acids Modulate ATPases Involved in Duodenal Ca Absorption. *Prostaglandins Leukot. Essent. Fat. Acids* **2003**, *68*, 423–429. [CrossRef]
11. Casado-Díaz, A.; Santiago-Mora, R.; Dorado, G.; Quesada-Gómez, J.M. The Omega-6 Arachidonic Fatty Acid, but Not the Omega-3 Fatty Acids, Inhibits Osteoblastogenesis and Induces Adipogenesis of Human Mesenchymal Stem Cells: Potential Implication in Osteoporosis. *Osteoporos. Int.* **2013**, *24*, 1647–1661. [CrossRef] [PubMed]
12. Coetzee, M.; Haag, M.; Joubert, A.M.; Kruger, M.C. Effects of Arachidonic Acid, Docosahexaenoic Acid and Prostaglandin E2 on Cell Proliferation and Morphology of MG-63 and MC3T3-E1 Osteoblast-like Cells. *Prostaglandins Leukot. Essent. Fat. Acids* **2007**, *76*, 35–45. [CrossRef]
13. Zwart, S.R.; Pierson, D.; Mehta, S.; Gonda, S.; Smith, S.M. Capacity of Omega-3 Fatty Acids or Eicosapentaenoic Acid to Counteract Weightlessness-Induced Bone Loss by Inhibiting NF-KB Activation: From Cells to Bed Rest to Astronauts. *J. Bone Miner. Res.* **2010**, *25*, 1049–1057. [CrossRef]
14. Benetou, V.; Orfanos, P.; Zylis, D.; Sieri, S.; Contiero, P.; Tumino, R.; Giurdanella, M.C.; Peeters, P.H.M.; Linseisen, J.; Nieters, A.; et al. Diet and Hip Fractures among Elderly Europeans in the EPIC Cohort. *Eur. J. Clin. Nutr.* **2011**, *65*, 132–139. [CrossRef]
15. Virtanen, J.K.; Mozaffarian, D.; Willett, W.C.; Feskanich, D. Dietary Intake of Polyunsaturated Fatty Acids and Risk of Hip Fracture in Men and Women. *Osteoporos. Int.* **2012**, *23*, 2615–2624. [CrossRef]
16. Bao, M.; Zhang, K.; Wei, Y.; Hua, W.; Gao, Y.; Li, X.; Ye, L. Therapeutic Potentials and Modulatory Mechanisms of Fatty Acids in Bone. *Cell Prolif.* **2020**, *53*, e12735. [CrossRef] [PubMed]
17. Orchard, T.S.; Cauley, J.A.; Frank, G.C.; Neuhouser, M.L.; Robinson, J.G.; Snetselaar, L.; Tylavsky, F.; Wactawski-Wende, J.; Young, A.M.; Lu, B.; et al. Fatty Acid Consumption and Risk of Fracture in the Women's Health Initiative. *Am. J. Clin. Nutr.* **2010**, *92*, 1452–1460. [CrossRef] [PubMed]
18. Orchard, T.S.; Pan, X.; Cheek, F.; Ing, S.W.; Jackson, R.D. A Systematic Review of Omega-3 Fatty Acids and Osteoporosis. *Br. J. Nutr.* **2012**, *107*, S253–S260. [CrossRef]
19. Farina, E.K.; Kiel, D.P.; Roubenoff, R.; Schaefer, E.J.; Cupples, L.A.; Tucker, K.L. Protective Effects of Fish Intake and Interactive Effects of Long-Chain Polyunsaturated Fatty Acid Intakes on Hip Bone Mineral Density in Older Adults: The Framingham Osteoporosis Study123. *Am. J. Clin. Nutr.* **2011**, *93*, 1142–1151. [CrossRef] [PubMed]
20. Lavado-García, J.; Roncero-Martin, R.; Moran, J.M.; Pedrera-Canal, M.; Aliaga, I.; Leal-Hernandez, O.; Rico-Martin, S.; Canal-Macias, M.L. Long-Chain Omega-3 Polyunsaturated Fatty Acid Dietary Intake is Positively Associated with Bone Mineral Density in Normal and Osteopenic Spanish Women. *PLoS ONE* **2018**, *13*, e0190539. [CrossRef]
21. Yuan, S.; Lemming, E.W.; Michaëlsson, K.; Larsson, S.C. Plasma Phospholipid Fatty Acids, Bone Mineral Density and Fracture Risk: Evidence from a Mendelian Randomization Study. *Clin. Nutr.* **2020**, *39*, 2180–2186. [CrossRef] [PubMed]
22. Takkunen, M.; Ågren, J.; Kuusisto, J.; Laakso, M.; Uusitupa, M.; Schwab, U. Dietary Fat in Relation to Erythrocyte Fatty Acid Composition in Men. *Lipids* **2013**, *48*, 1093–1102. [CrossRef]
23. Patel, P.S.; Sharp, S.J.; Jansen, E.; Luben, R.N.; Khaw, K.-T.; Wareham, N.J.; Forouhi, N.G. Fatty Acids Measured in Plasma and Erythrocyte-Membrane Phospholipids and Derived by Food-Frequency Questionnaire and the Risk of New-Onset Type 2 Diabetes: A Pilot Study in the European Prospective Investigation into Cancer and Nutrition (EPIC)–Norfolk Cohort. *Am. J. Clin. Nutr.* **2010**, *92*, 1214–1222. [CrossRef]
24. Sun, Q.; Ma, J.; Campos, H.; Hankinson, S.E.; Hu, F.B. Comparison between Plasma and Erythrocyte Fatty Acid Content as Biomarkers of Fatty Acid Intake in US Women. *Am. J. Clin. Nutr.* **2007**, *86*, 74–81. [CrossRef]
25. Astorg, P.; Bertrais, S.; Laporte, F.; Arnault, N.; Estaquio, C.; Galan, P.; Favier, A.; Hercberg, S. Plasma N-6 and *n*-3 Polyunsaturated Fatty Acids as Biomarkers of Their Dietary Intakes: A Cross-Sectional Study within a Cohort of Middle-Aged French Men and Women. *Eur. J. Clin. Nutr.* **2008**, *62*, 1155–1161. [CrossRef] [PubMed]
26. Lavado-Garcia, J.M.; Calderon-Garcia, J.F.; Moran, J.M.; Canal-Macias, M.L.; Rodriguez-Dominguez, T.; Pedrera-Zamorano, J.D. Bone Mass of Spanish School Children: Impact of Anthropometric, Dietary and Body Composition Factors. *J. Bone Miner. Metab.* **2012**, *30*, 193–201. [CrossRef]
27. Moreiras, O.; Carbajal, A.; Cabrera, L.; Cuadrado, C. *Tablas de Composición de Alimentos*, 16th ed.; Pirámide: Madrid, Spain, 2013.
28. Templeton, G. A Two-Step Approach for Transforming Continuous Variables to Normal: Implications and Recommendations for IS Research. *Commun. Assoc. Inf. Syst.* **2011**, *28*, 41–58. [CrossRef]
29. Svensson, M.; Schmidt, E.B.; Jørgensen, K.A.; Christensen, J.H. The Effect of N-3 Fatty Acids on Lipids and Lipoproteins in Patients Treated with Chronic Haemodialysis: A Randomized Placebo-Controlled Intervention Study. *Nephrol. Dial. Transplant.* **2008**, *23*, 2918–2924. [CrossRef] [PubMed]
30. Kuroda, T.; Ohta, H.; Onoe, Y.; Tsugawa, N.; Shiraki, M. Intake of Omega-3 Fatty Acids Contributes to Bone Mineral Density at the Hip in a Younger Japanese Female Population. *Osteoporos. Int.* **2017**, *28*, 2887–2891. [CrossRef] [PubMed]

31. Longo, A.B.; Ward, W.E. PUFAs, Bone Mineral Density, and Fragility Fracture: Findings from Human Studies. *Adv. Nutr.* **2016**, *7*, 299–312. [CrossRef]
32. Sahni, S.; Mangano, K.M.; McLean, R.R.; Hannan, M.T.; Kiel, D.P. Dietary Approaches for Bone Health: Lessons from the Framingham Osteoporosis Study. *Curr. Osteoporos. Rep.* **2015**, *13*, 245–255. [CrossRef]
33. Bellissimo, M.P.; Ziegler, T.R.; Jones, D.P.; Liu, K.H.; Fernandes, J.; Roberts, J.L.; Weitzmann, M.N.; Pacifici, R.; Alvarez, J.A. Plasma High-Resolution Metabolomics Identifies Linoleic Acid and Linked Metabolic Pathways Associated with Bone Mineral Density. *Clin. Nutr.* **2020**. [CrossRef]
34. Calderon-Garcia, J.F.; Moran, J.M.; Roncero-Martin, R.; Rey-Sanchez, P.; Rodriguez-Velasco, F.J.; Pedrera-Zamorano, J.D. Dietary Habits, Nutrients and Bone Mass in Spanish Premenopausal Women: The Contribution of Fish to Better Bone Health. *Nutrients* **2012**, *5*, 10–22. [CrossRef] [PubMed]
35. Harris, T.B.; Song, X.; Reinders, I.; Lang, T.F.; Garcia, M.E.; Siggeirsdottir, K.; Sigurdsson, S.; Gudnason, V.; Eiriksdottir, G.; Sigurdsson, G.; et al. Plasma Phospholipid Fatty Acids and Fish-Oil Consumption in Relation to Osteoporotic Fracture Risk in Older Adults: The Age, Gene/Environment Susceptibility Study. *Am. J. Clin. Nutr.* **2015**, *101*, 947–955. [CrossRef] [PubMed]
36. Chen, J.S.; Hill, C.L.; Lester, S.; Ruediger, C.D.; Battersby, R.; Jones, G.; Cleland, L.G.; March, L.M. Supplementation with Omega-3 Fish Oil Has No Effect on Bone Mineral Density in Adults with Knee Osteoarthritis: A 2-Year Randomized Controlled Trial. *Osteoporos. Int.* **2016**, *27*, 1897–1905. [CrossRef] [PubMed]
37. Watkins, B.A.; Li, Y.; Seifert, M.F. Dietary Ratio of N-6/*n*-3 PUFAs and Docosahexaenoic Acid: Actions on Bone Mineral and Serum Biomarkers in Ovariectomized Rats. *J. Nutr. Biochem.* **2006**, *17*, 282–289. [CrossRef]
38. Simopoulos, A.P. The Importance of the Omega-6/Omega-3 Fatty Acid Ratio in Cardiovascular Disease and Other Chronic Diseases. *Exp. Biol. Med.* **2008**, *233*, 674–688. [CrossRef] [PubMed]
39. Orchard, T.S.; Ing, S.W.; Lu, B.; Belury, M.A.; Johnson, K.; Wactawski-Wende, J.; Jackson, R.D. The Association of Red Blood Cell N-3 and *n*-6 Fatty Acids with Bone Mineral Density and Hip Fracture Risk in the Women's Health Initiative. *J. Bone Miner. Res.* **2013**, *28*, 505–515. [CrossRef]
40. Moon, H.-J.; Kim, T.-H.; Byun, D.-W.; Park, Y. Positive Correlation between Erythrocyte Levels of N-3 Polyunsaturated Fatty Acids and Bone Mass in Postmenopausal Korean Women with Osteoporosis. *Ann. Nutr. Metab.* **2012**, *60*, 146–153. [CrossRef]
41. Mei, Z.; Dong, X.; Qian, Y.; Hong, D.; Xie, Z.; Yao, G.; Qin, A.; Gao, S.; Hu, J.; Liang, L.; et al. Association between the Metabolome and Bone Mineral Density in a Chinese Population. *EBioMedicine* **2020**, *62*. [CrossRef]
42. Rozner, R.; Vernikov, J.; Griess-Fishheimer, S.; Travinsky, T.; Penn, S.; Schwartz, B.; Mesilati-Stahy, R.; Argov-Argaman, N.; Shahar, R.; Monsonego-Ornan, E. The Role of Omega-3 Polyunsaturated Fatty Acids from Different Sources in Bone Development. *Nutrients* **2020**, *12*, 3494. [CrossRef]
43. Abou-Saleh, H.; Ouhtit, A.; Halade, G.V.; Rahman, M.M. Bone Benefits of Fish Oil Supplementation Depend on Its EPA and DHA Content. *Nutrients* **2019**, *11*, 2701. [CrossRef] [PubMed]
44. Watkins, B.A.; Li, Y.; Seifert, M.F. Nutraceutical Fatty Acids as Biochemical and Molecular Modulators of Skeletal Biology. *J. Am. Coll. Nutr.* **2001**, *20*, 410S–416S, discussion 417S–420S. [CrossRef]
45. Watkins, B.A.; Lippman, H.E.; Le Bouteiller, L.; Li, Y.; Seifert, M.F. Bioactive Fatty Acids: Role in Bone Biology and Bone Cell Function. *Prog. Lipid. Res.* **2001**, *40*, 125–148. [CrossRef]
46. Loi, F.; Córdova, L.A.; Pajarinen, J.; Lin, T.; Yao, Z.; Goodman, S.B. Inflammation, Fracture and Bone Repair. *Bone* **2016**, *86*, 119–130. [CrossRef] [PubMed]
47. Li, M.; Ke, H.Z.; Qi, H.; Healy, D.R.; Li, Y.; Crawford, D.T.; Paralkar, V.M.; Owen, T.A.; Cameron, K.O.; Lefker, B.A.; et al. A Novel, Non-Prostanoid EP2 Receptor-Selective Prostaglandin E2 Agonist Stimulates Local Bone Formation and Enhances Fracture Healing. *J. Bone Miner. Res.* **2003**, *18*, 2033–2042. [CrossRef] [PubMed]
48. Yang, S.; Feskanich, D.; Willett, W.C.; Eliassen, A.H.; Wu, T. Association between Global Biomarkers of Oxidative Stress and Hip Fracture in Postmenopausal Women: A Prospective Study. *J. Bone Miner. Res.* **2014**, *29*, 2577–2583. [CrossRef] [PubMed]
49. Manolagas, S.C.; Parfitt, A.M. What Old Means to Bone. *Trends. Endocrinol. Metab.* **2010**, *21*, 369–374. [CrossRef] [PubMed]
50. Zhao, G.; Etherton, T.D.; Martin, K.R.; Gillies, P.J.; West, S.G.; Kris-Etherton, P.M. Dietary Alpha-Linolenic Acid Inhibits Proinflammatory Cytokine Production by Peripheral Blood Mononuclear Cells in Hypercholesterolemic Subjects. *Am. J. Clin. Nutr.* **2007**, *85*, 385–391. [CrossRef]
51. Larsson, S.C.; Kumlin, M.; Ingelman-Sundberg, M.; Wolk, A. Dietary Long-Chain *n*-3 Fatty Acids for the Prevention of Cancer: A Review of Potential Mechanisms. *Am. J. Clin. Nutr.* **2004**, *79*, 935–945. [CrossRef] [PubMed]
52. Sun, D.; Krishnan, A.; Zaman, K.; Lawrence, R.; Bhattacharya, A.; Fernandes, G. Dietary N-3 Fatty Acids Decrease Osteoclastogenesis and Loss of Bone Mass in Ovariectomized Mice. *J. Bone Miner. Res.* **2003**, *18*, 1206–1216. [CrossRef] [PubMed]
53. Akasaka, H.; Ruan, K.-H. Identification of the Two-Phase Mechanism of Arachidonic Acid Regulating Inflammatory Prostaglandin E2 Biosynthesis by Targeting COX-2 and MPGES-1. *Arch. Biochem. Biophys.* **2016**, *603*, 29–37. [CrossRef]
54. Coetzee, M.; Haag, M.; Claassen, N.; Kruger, M.C. Stimulation of Prostaglandin E2 (PGE2) Production by Arachidonic Acid, Oestrogen and Parathyroid Hormone in MG-63 and MC3T3-E1 Osteoblast-like Cells. *Prostaglandins Leukot. Essent. Fat. Acids* **2005**, *73*, 423–430. [CrossRef]
55. Kanematsu, M.; Sato, T.; Takai, H.; Watanabe, K.; Ikeda, K.; Yamada, Y. Prostaglandin E2 Induces Expression of Receptor Activator of Nuclear Factor-Kappa B Ligand/Osteoprotegrin Ligand on Pre-B Cells: Implications for Accelerated Osteoclastogenesis in Estrogen Deficiency. *J. Bone Miner. Res.* **2000**, *15*, 1321–1329. [CrossRef] [PubMed]

56. Suda, K.; Udagawa, N.; Sato, N.; Takami, M.; Itoh, K.; Woo, J.-T.; Takahashi, N.; Nagai, K. Suppression of Osteoprotegerin Expression by Prostaglandin E2 Is Crucially Involved in Lipopolysaccharide-Induced Osteoclast Formation. *J. Immunol.* **2004**, *172*, 2504–2510. [CrossRef] [PubMed]
57. Wennberg, M.; Tornevi, A.; Johansson, I.; Hörnell, A.; Norberg, M.; Bergdahl, I.A. Diet and Lifestyle Factors Associated with Fish Consumption in Men and Women: A Study of Whether Gender Differences Can Result in Gender-Specific Confounding. *Nutr. J.* **2012**, *11*, 101. [CrossRef]
58. Galli, C.; Trzeciak, H.I.; Paoletti, R. Effects of Dietary Fatty Acids on the Fatty Acid Composition of Brain Ethanolamine Phosphoglyceride: Reciprocal Replacement of *n*-6 and *n*-3 Polyunsaturated Fatty Acids. *Biochim. Biophys. Acta Lipids Lipid Metab.* **1971**, *248*, 449–454. [CrossRef]

Article

Association of Dietary Total Antioxidant Capacity with Bone Mass and Osteoporosis Risk in Korean Women: Analysis of the Korea National Health and Nutrition Examination Survey 2008–2011

Donghyun Kim [1,†], Anna Han [2,†] and Yongsoon Park [1,*]

[1] Department of Food and Nutrition, Hanyang University, 222 Wangsimni-ro, Seongdong-gu, Seoul 04763, Korea; apoqalik@hanyang.ac.kr

[2] Department of Cancer Biology, Thomas Jefferson University, Philadelphia, PA 19107, USA; annahan8659@gmail.com

* Correspondence: yongsoon@hanyang.ac.kr; Tel.: +82-2-2220-1205

† These authors contributed equally to this work.

Abstract: Antioxidant intake has been suggested to be associated with a reduced osteoporosis risk, but the effect of dietary total antioxidant capacity (TAC) on bone health and the risk of osteoporosis remains unclear. We aimed to assess the hypothesis that dietary TAC is positively associated with bone mass and negatively related to the risk of osteoporosis in Korean women. This cross-sectional study was performed using data from the Korea National Health and Nutrition Examination Survey. Dietary TAC was estimated using task automation and an algorithm with 24-h recall data. In total, 8230 pre- and postmenopausal women were divided into four groups according to quartiles of dietary TAC. Dietary TAC was negatively associated with the risk of osteoporosis (odds ratio, 0.73; 95% confidence interval, 0.54–0.99; *p*-value = 0.045) in postmenopausal women, but not in premenopausal women. Dietary TAC was positively associated with bone mineral content (BMC) and bone mineral density of the femoral neck and lumbar spine in postmenopausal women and BMC of the total femur and lumbar spine in premenopausal women. Our study suggests that dietary TAC is inversely associated with the risk of osteoporosis in postmenopausal women and positively associated with bone mass in both pre- and postmenopausal women.

Keywords: bone mass; dietary total antioxidant capacity; KNHANES; menopause; osteoporosis

Citation: Kim, D.; Han, A.; Park, Y. Association of Dietary Total Antioxidant Capacity with Bone Mass and Osteoporosis Risk in Korean Women: Analysis of the Korea National Health and Nutrition Examination Survey 2008–2011. *Nutrients* **2021**, *13*, 1149. https://doi.org/10.3390/nu13041149

Academic Editor: Jose M. Moran

Received: 4 March 2021
Accepted: 27 March 2021
Published: 31 March 2021

Publisher's Note: MDPI stays neutral with regard to jurisdictional claims in published maps and institutional affiliations.

1. Introduction

Osteoporosis is characterized by impairments in bone strength, elevating the risk of skeletal fractures, and is a major public health concern [1]. According to the Korea National Health and Nutrition Examination Survey (KNHANES) 2008–2011, the prevalence of osteoporosis in Korea was reported to be 17.2% in adult women and 38% in postmenopausal women ($\geq$50 years old), indicating that menopause is an important risk factor [2]. Furthermore, the prevalence of osteoporosis in Korea has also gradually increased due to the aging population, sedentary lifestyle, and westernized dietary patterns [2,3].

Oxidative stress increases bone remodeling turnover and bone mass loss by increasing the inflammatory response and modulating the survival and differentiation of osteoblasts, whereas antioxidants inhibit oxidative stress and consequently prevent the loss of bone mass [4]. A recent study demonstrated that the intake of fruits and vegetables containing abundant antioxidants was associated with reduced bone mass loss and a reduced risk of osteoporosis in postmenopausal women [5]. Although epidemiological studies have reported a positive association between bone mineral density (BMD) and the intake of single antioxidants, such as vitamin C, vitamin E, carotenoids, and flavonoids in both pre- menopausal and postmenopausal women [6–11], the beneficial effects of antioxidant intake on bone health have been controversial. In fact, the consumption of vitamin C has been found to be positively correlated with the risk of osteoporosis in several studies [12–14], but not in others [15,16].

Interestingly, the beneficial effects of antioxidants on bone mass and the risk of osteoporosis differ depending on the anatomical site of the bones [6,8,17]. In postmenopausal women, β-carotene consumption was positively associated with femoral neck, total hip, and whole-body BMD, and the highest quintile of β-carotene consumption was associated with a decreased risk of osteopenia at the lumbar spine [6]. In contrast, increased intake of β-cryptoxanthin led to a lower risk of osteopenia at the total hip and lumbar spine, although it showed a positive correlation only with total hip BMD [6].

Previous inconsistent observations regarding the association of the intake of single antioxidants with bone mass and osteoporosis risk can be partially explained by the fact that the diet contains various antioxidants that exert cumulative and synergistic effects. Recently, dietary total antioxidant capacity (TAC) has been found to be a practical indicator to evaluate the overall antioxidant capacity of foods [18]. Rather than a simple sum of single dietary antioxidants, dietary TAC provides comprehensive information about the accumulated antioxidant capacities of different diets; therefore, dietary TAC has been applied in many studies to assess the precise preventive effects of antioxidants on diseases, including inflammatory diseases, cardiovascular diseases, and cancers [19–21]. However, the association between dietary TAC and the risk of osteoporosis has not yet been studied. The present study hypothesized that dietary TAC is positively associated with bone mass and negatively associated with the risk of osteoporosis in premenopausal and postmenopausal women, and analyses were performed using data from the KNHANES (2008–2011).

2. Materials and Methods

2.1. Study Population

Participants ($n = 37{,}753$) from the KNHANES (2008–2011) were included in this study. The KNHANES is a cross-sectional nationwide survey designed to evaluate the nutritional and health status of non-institutionalized and civilian Korean populations. The KNHANES consists of health interviews, health examinations, and nutrition surveys, and the survey uses a multistage clustered probability sampling design to select a nationally representative sample. The Institutional Review Board of the Korea National Statistical Office, Centers for Disease Control and Prevention approved the KNHANES protocol. Informed consent was obtained from all participants [22]. This study was approved by the Hanyang University Institutional Review Board (No. HYUIRB-202101-007).

The exclusion criteria for this study were as follows: male sex ($n = 17{,}195$); age < 19 years ($n = 4479$); missing data on bone mineral content (BMC) or BMD ($n = 4870$); missing data on baseline characteristics ($n = 1395$); Z-scores of dietary TAC, BMD, and BMC greater than +3.29 or less than −3.29 ($n = 251$); extreme energy intake (<500 kcal/day or >4000 kcal/day; $n = 250$); renal failure ($n = 19$); pregnancy or lactation ($n = 148$); hormone replacement therapy ($n = 684$); and osteoporosis treatment ($n = 232$). A total of 8230 female participants were included in the final analyses.

2.2. Study Variables

A health interview questionnaire was used to obtain data on age, sex, postmenopausal status, smoking status, alcohol consumption, and exercise status. Postmenopausal status was defined on the basis of a self-reported questionnaire survey containing questions regarding whether one year had passed since the last menstruation or hysterectomy. Waist circumference (WC) and body mass index (BMI) were measured as indices of body measurements. Regular exercise was defined as walking for more than 30 min at least five times per week or vigorous physical activity for at least 20 min per day, three days per week. For biochemical analysis, a fasting blood sample was taken, refrigerated, and analyzed within 24 h at the Central Testing Center in Seoul, Korea. The serum concentration of

25 hydroxyvitamin D (25(OH)D) was measured using the radioimmunoassay method (1470 WIZARD Gamma-Counter, PerkinElmer, Finland).

Dietary intake data were collected using one day of the 24-h recall method during household interviews. Daily intakes of energy, protein, calcium, phosphorus, potassium, and sodium were calculated using the Korean Food Composition Table [23].

BMC and BMD at the lumbar spine (L1–L4), total femur, and femoral neck were measured by trained technicians using dual-energy X-ray absorptiometry (DISCOVERY-W fan-beam densitometer; HOLOGIC, MA, USA). Osteopenia or osteoporosis was diagnosed using the minimum T-score of BMD in three sites (lumbar spine, total femur, and femoral neck) based on the criteria of the KNHANES [24]. Briefly, the T-scores of the lumbar spine, total femur, and femoral neck were classified as follows: greater than or equal to -1, normal; less than -1 to more than -2.5, osteopenia; and less than or equal to -2.5, osteoporosis.

2.3. Development of the TAC Database and Estimation of Dietary TAC

According to Floegel et al. [18], the theoretical TAC value of each food item was obtained by summing the TAC values of individual antioxidants derived by multiplying the contents of 42 individual antioxidants in the food by their antioxidant capacities; antioxidant capacities were determined by the 2,2-azino-bis-3-ethylbenzthiazoline-6-sulphonic acid assay and expressed as vitamin C equivalents (VCE). The individual antioxidants were β-carotene, α-carotene, β-cryptoxanthin, lycopene, lutein, zeaxanthin, retinol, ascorbic acid, α-tocopherol, γ-tocopherol, quercetin, kaempferol, myricetin, isorhamnetin, luteolin, apigenin, hesperetin, naringenin, eriodictyol, (+)-catechin, (+)-gallocatechin, (–)-epicatechin, (–)-epigallocatechin, (–)-epicatechin 3-gallate, (–)-epigallocatechin 3-gallate, theaflavin, theaflavin 3-gallate, theaflavin 3′-gallate, theaflavin 3,3′-digallate, cyanidin, delphinidin, malvidin, pelargonidin, peonidin, petunidin, daidzein, genistein, glycitein, biochanin A, formononetin, procyanidin, and cinnamtannin B1.

The contents of 42 individual antioxidants were collected from databases produced by the Korea Rural Development Administration (RDA) [25,26] and the United States Department of Agriculture [27–29]. When two or more values for an individual food item were available from these databases, the RDA value was prioritized. For food items without available values in these government publications, we used the values from published papers [30–33]. The database was expanded by estimating values based on similar food items and moisture conversion factors suggested in the KNHANES, and by applying logical zeros. The missing values were denoted as zeros [34]. The database covered 95.34% of the food intake.

Dietary TAC was estimated by creating a program linking food consumption data from the KNHANES with the TAC database. The dietary TAC (mg VCE) of individual food items was determined by multiplying the daily consumption of the item (g) by the theoretical TAC of each food (mg VCE/100 g) in our TAC database (Equation (1)). Daily TAC from the diet was the sum of daily TAC from all food items reported in the one day of 24-h recall (Equation (2)). This series of operations was performed efficiently through computational automation (Supplement S1).

Equation (1):

$$Theoretical\ TAC = \sum \left(antioxidant\ content \frac{mg}{100g} \times antioxidant\ capacity \frac{mg\ VCE}{100g} \right) \quad (1)$$

Equation (2):

$$Daily\ dietary\ TAC = \sum (Daily\ intake\ food\ amount \times theoretical\ TAC\ (per\ 1g\ of\ food\ )) \quad (2)$$

2.4. Statistical Analyses

All statistical analyses were performed using complex sample survey data in SPSS version 26.0 (SPSS Inc., Chicago, IL, USA). Sample weights obtained from the KNHANES were used to obtain unbiased estimates of means and frequencies that were nationally representative of the Korean population [35]. Continuous variables were expressed as the mean ± standard error of the mean, and categorical variables were expressed as frequencies and percentages. The characteristics of and risk factors for osteoporosis were compared between the groups using the Student's *t*-test for continuous variables and the chi-square test for categorical variables. The values of dietary TAC, BMC, or BMD at the total femur, femoral neck, and lumbar spine were standardized to identify univariate outliers.

Multiple regression analyses were performed to screen for unsuitable potential covariates and to assess the relationships between dietary TAC and BMC or BMD, considering the suitable potential covariates. In multivariate models, covariates showing a p-value < 0.20 were selected as confounding factors and included in the adjusted model [36]. All groups were subdivided into four groups according to quartiles of dietary TAC. Analysis of covariance (ANCOVA) with Bonferroni correction was performed to assess mean differences in BMC or BMD among the quartiles after adjustment for confounding variables. A multivariable logistic regression model was used to examine the association between dietary TAC and the risk of osteoporosis. The p-value for the trend was calculated by employing multivariate logistic regression analyses and by handling the median value of each dietary TAC quartile as a continuous value. Statistical significance was set at a p-value < 0.05.

3. Results

3.1. Study Design and Baseline Characteristics of Participants

Figure 1 outlines the study design and selection of participants for the present study. Table 1 presents the participants' baseline characteristics and the risk factors of osteoporosis in the study population. Both postmenopausal and premenopausal women with osteoporosis had a significantly lower WC, BMI, BMC, and BMD than those without osteoporosis. In postmenopausal women, participants with osteoporosis were older, had a higher smoking frequency, had lower alcohol consumption, and engaged in more regular patterns of exercise than those without osteoporosis; additionally, they also had lower dietary intakes of energy, protein, calcium, phosphorous, potassium, and sodium than those without osteoporosis. Premenopausal women with osteoporosis had markedly lower 25(OH)D levels than those without osteoporosis.

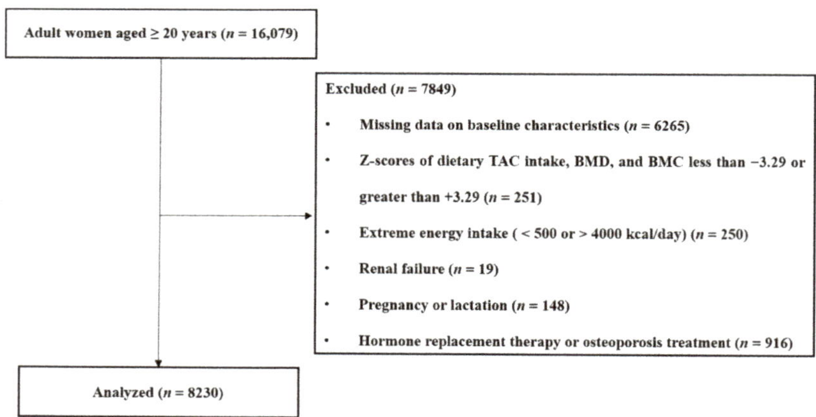

Figure 1. Summary of participant selection in the present study. TAC, total antioxidant capacity; BMC, bone mineral content; BMD, bone mineral density.

Table 1. Baseline characteristics of the participants and risk factors of osteoporosis in the study population.

Variables	Postmenopausal Women			Premenopausal Women		
	Non-Osteoporosis (*n* = 2260)	Osteoporosis (*n* = 1299)	*p*-Value	Non-Osteoporosis (*n* = 4606)	Osteoporosis (*n* = 65)	*p*-Value
Age (years)	60.46 ± 0.19	69.27 ± 0.22	<0.001	36.70 ± 0.13	37.69 ± 1.14	0.670
WC (cm)	83.82 ± 0.19	80.77 ± 0.27	<0.001	75.34 ± 0.14	68.80 ± 0.81	<0.001
BMI (kg/m^2)	24.81 ± 0.07	23.28 ± 0.09	<0.001	22.65 ± 0.05	19.98 ± 0.30	<0.001
Smoking status, *n* (%)			0.002			0.443
Never	2100 (93.7)	1149 (89.0)		4122 (89.8)	57 (87.7)	
Former	56 (2.5)	61 (4.7)		184 (4.0)	1 (1.5)	
Current	86 (3.8)	81 (6.3)		285 (6.2)	7 (10.8)	
Alcohol consumption, *n* (%)	627 (28.0)	246 (19.1)	<0.001	2330 (50.8)	26 (41.3)	0.162
Regular exercise, *n* (%)	557 (24.9)	219 (17.0)	<0.001	1034 (22.6)	13 (20.0)	0.367
25(OH)D (ng/mL)	18.66 ± 0.15	18.42 ± 0.20	0.116	16.05 ± 0.08	14.11 ± 0.73	0.047
Dietary intake						
Energy intake (kcal/day)	1587.56 ± 11.43	1449.32 ± 13.78	<0.001	1694.50 ± 8.77	1686.48 ± 67.41	0.517
Protein (g/day)	53.81 ± 0.551	45.87 ± 0.663	<0.001	61.93 ± 0.411	62.19 ± 3.126	0.915
Calcium (mg/day)	435.70 ± 6.43	357.66 ± 12.24	<0.001	457.82 ± 4.15	474.80 ± 32.74	0.491
Phosphorus (mg/day)	974.54 ± 8.54	851.73 ± 10.22	<0.001	1025.90 ± 5.96	1067.54 ± 47.65	0.472
Potassium (mg/day)	2695.38 ± 29.90	2245.41 ± 35.53	<0.001	2728.75 ± 18.57	2826.00 ± 162.31	0.468
Sodium (mg/day)	3913.31 ± 53.89	3513.67 ± 70.59	<0.001	4344.46 ± 42.51	4773.58 ± 384.46	0.302
BMC (g)						
Lumbar spine	48.87 ± 0.233	34.73 ± 0.239	<0.001	55.92 ± 0.155	41.27 ± 0.729	<0.001
Total femur	28.22 ± 0.092	22.76 ± 0.104	<0.001	29.21 ± 0.064	21.95 ± 0.387	<0.001
Femoral neck	3.34 ± 0.010	2.59 ± 0.011	<0.001	3.61 ± 0.008	2.70 ± 0.041	<0.001
BMD (g/cm^2)						
Lumbar spine	0.87 ± 0.002	0.67 ± 0.003	<0.001	0.95 ± 0.002	0.74 ± 0.010	<0.001
Total femur	0.83 ± 0.002	0.67 ± 0.002	<0.001	0.88 ± 0.002	0.70 ± 0.009	<0.001
Femoral neck	0.68 ± 0.002	0.53 ± 0.002	<0.001	0.74 ± 0.001	0.56 ± 0.007	<0.001

Values are expressed as the mean $\pm$ standard error of the mean for continuous variables or as the number (percentage) for categorical variables. WC, waist circumference; BMI, body mass index; BMC, bone mineral content; BMD, bone mineral density.

3.2. Associations between Dietary TAC and Bone Health

Logistic regression analysis showed that dietary TAC was negatively associated with the risk of osteoporosis before and after adjustment for potential confounders in postmenopausal women, but not in premenopausal women (Table 2). Postmenopausal women consuming food items corresponding to a daily dietary TAC of $\geq$456.89 mg VCE/day had a low risk of osteoporosis. Furthermore, there were significant positive correlations between dietary TAC, BMC, and BMD in postmenopausal women (Table 3). Dietary TAC was positively correlated only with the BMC of the total femur and lumbar spine in premenopausal women.

Lastly, ANCOVA showed that dietary TAC was positively associated with the BMC of the lumbar spine, total femur, and femoral neck and the BMD of the lumbar spine and femoral neck before and after adjustment for potential confounders in postmenopausal women (Table 4). In premenopausal women, there was a significant and positive association between dietary TAC and the BMC of the total femur and lumbar spine (Table 4).

Table 2. Associations between dietary total antioxidant capacity (TAC) and the risk of osteoporosis in the study population.

Variables	Quartiles of Dietary TAC (mg VCE/day)				*p*-Value
	Q1	Q2	Q3	Q4	
Postmenopausal women, *n* (osteoporosis %)	890 (47.0)	890 (37.5)	890 (33.4)	889 (28.1)	
Cut off (mg VCE/day)	<150.96	150.96 ≤ to < 267.65	267.65 ≤ to < 456.89	≥456.89	
Median dietary TAC (mg VCE/day)	93.48	208.26	347.32	643.52	
Unadjusted OR (95% CI)	1	0.673 (0.525–0.863)	0.618 (0.486–0.787)	0.398 (0.315–0.504)	<0.001
Adjusted OR (95% CI)	1	0.952 (0.715–0.268)	1.066 (0.805–1.410)	0.732 (0.540–0.992)	0.045
Premenopausal women, *n* (osteoporosis %)	1167 (1.7)	1168 (1.0)	1168 (1.4)	1168 (1.5)	
Cut off (mg VCE/day)	<169.28	169.28 ≤ to < 286.79	286.79 ≤ to < 474.48	≥474.48	
Median dietary TAC (mg VCE/day)	109.36	226.41	361.06	666.44	
Unadjusted OR (95% CI)	1	0.515 (0.216–1.228)	0.708 (0.337–1.487)	1.056 (0.489–2.283)	0.590
Adjusted OR (95% CI)	1	0.554 (0.232–1.318)	0.825 (0.395–1.723)	1.220 (0.555–2.681)	0.394

Odds ratios (ORs) and 95% confidence intervals (CIs) are presented. The logistic regression model was adjusted for age, waist circumference, energy intake, sodium intake, smoking status, and regular exercise for postmenopausal women, and adjusted for waist circumference and 25(OH)D levels for premenopausal women. VCE, vitamin C equivalents.

Table 3. Correlation of dietary total antioxidant capacity (TAC) with bone mass.

Variables	Postmenopausal Women		Premenopausal Women	
	r [1]	*p*-Value	*r* [1]	*p*-Value
BMC (g)				
Lumbar spine	0.146	<0.001	0.043	0.005
Total femur	0.153	<0.001	0.034	0.037
Femoral neck	0.179	<0.001	0.008	0.646
BMD (g/cm^2)				
Lumbar spine	0.144	<0.001	0.023	0.154
Total femur	0.170	<0.001	0.014	0.395
Femoral neck	0.177	<0.001	0.012	0.480

[1] All values represent correlations (r). BMC, bone mineral content; BMD, bone mineral density.

Table 4. Bone mass according to quartiles of dietary total antioxidant capacity (TAC) in the study population.

Variables	Quartiles of Dietary TAC (mg VCE/day)				*p*-Value [a]	*p*-Value [b]
	Q1	Q2	Q3	Q4		
Postmenopausal women, *n*	890	890	890	889		
Dietary TAC (mg VCE/day)	90.10 ± 1.32	207.92 ± 1.14	352.61 ± 1.81	726.45 ± 8.54		
BMC (g)						
Lumbar spine	40.86 ± 0.409	43.50 ± 0.421	44.07 ± 0.411	45.94 ± 0.409	<0.001	0.049
Total femur	25.11 ± 0.166	26.01 ± 0.163	26.61 ± 0.158	27.19 ± 0.166	<0.001	0.030
Femoral neck	2.89 ± 0.019	3.05 ± 0.019	3.11 ± 0.019	3.21 ± 0.019	<0.001	0.016
BMD (g/cm^2)						
Lumbar spine	0.77 ± 0.005	0.80 ± 0.005	0.81 ± 0.005	0.82 ± 0.005	<0.001	0.017
Total femur	0.74 ± 0.004	0.77 ± 0.004	0.78 ± 0.004	0.80 ± 0.004	<0.001	0.080
Femoral neck	0.59 ± 0.004	0.62 ± 0.004	0.63 ± 0.004	0.65 ± 0.004	<0.001	0.025
Premenopausal women, *n*	1167	1168	1168	1168		
Dietary TAC (mg VCE/day)	105.70 ± 1.16	226.70 ± 1.00	369.13 ± 1.60	762.04 ± 8.00		
BMC (g)						
Lumbar spine	58.42 ± 0.283	59.10 ± 0.283	59.49 ± 0.282	60.03 ± 0.297	0.001	0.002
Total femur	29.21 ± 0.137	29.44 ± 0.139	29.85 ± 0.142	29.92 ± 0.141	0.006	0.015
Femoral neck	3.72 ± 0.017	3.71 ± 0.016	3.76 ± 0.016	3.76 ± 0.017	0.527	0.850
BMD (g/cm^2)						
Lumbar spine	0.98 ± 0.003	0.99 ± 0.003	0.99 ± 0.003	0.99 ± 0.004	0.029	0.050
Total femur	0.90 ± 0.003	0.90 ± 0.003	0.91 ± 0.003	0.91 ± 0.003	0.093	0.231
Femoral neck	0.76 ± 0.003	0.76 ± 0.003	0.77 ± 0.003	0.77 ± 0.003	0.849	0.856

All values are presented as the mean ± standard error of the mean or the number of participants, as appropriate. [a] Unadjusted p-value for the differences in BMC and BMD according to quartiles of dietary TAC; [b] p-value for the differences in BMC and BMD according to quartiles of dietary TAC after adjustment for confounding factors. The covariates were age, waist circumference, energy intake, sodium intake, smoking status, and regular exercise for postmenopausal women, and waist circumference and 25(OH)D levels for premenopausal women. BMC, bone mineral content; BMD, bone mineral density.

4. Discussion

The present study showed that dietary TAC was positively associated with BMC and BMD and inversely associated with the risk of osteoporosis in postmenopausal women. This study also found a positive correlation between dietary TAC and the BMC of the lumbar spine and total femur in premenopausal women.

The intake of fruits and vegetables has a beneficial effect on bone mass in post-menopausal women [4,37]. Antioxidants present in fruits and vegetables have been highlighted as one of the key causes of beneficial outcomes for bone health [5]. Epidemiological studies have shown that BMD is positively related to the intake of vitamin C [38], vitamin E [8], carotenoids [6], and flavonoids [10,11] in postmenopausal women. Collectively, these observations indicated that the BMD of the femoral neck was strongly affected by the intake of antioxidants, while the BMD of the total hip and lumbar spine was slightly affected in postmenopausal women. In the present study, dietary TAC was significantly associated with the BMD of the femoral neck and lumbar spine, but not with the BMD of the total femur in postmenopausal women. It has been reported that the accumulation of yellow bone marrow due to aging is known to exert an effect related to an increase of inflammatory substances and inhibition of bone formation, and, as a result, inhibit skeletal homeostasis and cause a reduction in BMD [39,40], particularly femur BMD [41]. Thus, the confounding effects of variables such as age must be considered for the precise assessment of the effect of dietary TAC on femur BMD in postmenopausal women.

Furthermore, Li et al. [37] reported that the intake of fruits and vegetables was significantly associated with whole-body BMC, but not with the BMC and BMD of specific sites in premenopausal women. In premenopausal women, vitamin C intake was positively associated with femur BMD [42], vitamin E intake was positively correlated with lumbar BMD [9], and carotenoid intake was positively linked to the BMD of the total hip [6]. These findings suggest that the effect of a single antioxidant on BMD could be site-specific in premenopausal women. The present study also showed that dietary TAC was positively associated with the BMC of the lumbar spine and total femur in premenopausal women. In premenopausal women, bone loss caused by oxidative stress could be more prominent in the trabecular bone than in the cortical bone [5], therefore, sites with a greater proportion of trabecular bone, such as the lumbar spine and total femur, were more sensitive to antioxidant effects than the femoral neck [43].

The inverse association between the intake of fruit and vegetables and the risk of osteoporosis in postmenopausal women has been demonstrated previously [4,5,44], although some studies have shown controversial results regarding the association between the intake of single antioxidants and osteoporosis risk. Cross-sectional studies showed a significant negative association between the risk of osteoporosis and vitamin C intake in postmenopausal women (n = 1196 [12] and n = 1878 [13]) and men and women aged $\geq$50 years (n = 3047) [14]. In contrast, a cohort study reported that vitamin C and vitamin E intake was not related to the risk of osteoporosis in postmenopausal women (n = 187) [16]; a case-control study also reported that vitamin C intake was not associated with osteoporosis risk in postmenopausal women (n = 144) [15]. The small sample sizes in these studies might have reduced the accuracy of the statistical analyses [45], implying that the inconsistencies in the above findings might be due to small sample sizes. Our study analyzed a relatively large sample population (n = 3559) and demonstrated an inverse association between dietary TAC and the risk of osteoporosis in postmenopausal women.

The correlation between the risk of osteoporosis and the intake of fruits, vegetables, or antioxidants in premenopausal women has been highlighted and studied to a lesser extent, because estrogen prevents bone loss by maintaining bone homeostasis and lowering the risk of osteoporosis in premenopausal women [46]. The present study found no association between the risk of osteoporosis and dietary TAC in premenopausal women that might result from a very low incidence of osteoporosis in premenopausal women (1.0–1.7%) in the current study.

Independent with BMD, several biomarkers have been addressed to evaluate bone health and the risk of osteoporosis and osteoporotic fractures. Advanced glycation end products (AGEs), such as pentosidine, were associated with vertebral fracture, but not with lumbar BMD. [47]. Imbalance of serum homocysteine (Hcy) level has also been reported as a pathological biomarker of osteoporotic fractures [48]. Thus, it will be interesting to study the effects of a high dietary TAC diet on serum AGEs and Hcy levels, as well as the risk of osteoporotic fractures.

Consistent with our results, a vegan diet has been suggested to be associated with better bone health compared to non-vegans due to the inverse correlation between fruit and vegetable consumption and bone health [5,37]. However, a recent study found that vegetarians and vegans showed a lower BMD at the femoral neck and lumbar spine compared to non-vegans [49], which might come from the insufficient intake of dietary calcium, vitamin D, and high biological value proteins in the vegan diet [50]. The present study also observed that postmenopausal women with osteoporosis had a lower intake of calcium, phosphorous, protein, and energy intake, and premenopausal women with osteoporosis had a reduced level of 25(OH)D levels than those without osteoporosis. Since the original survey was a cross-sectional study, a cause-and-effect association between the lower nutrient levels and the risk of osteoporosis was unclear. However, high dietary TAC was still significantly inversely associated with the risk of osteoporosis in postmenopausal women and positively correlated to the site-specific BMC and BMD in both pre-and post-menopausal women after adjusting for energy intake. These observations suggest that dietary TAC might also need to be considered for the evaluation of the inverse correlation of a vegan diet and the risk of osteoporosis and bone health in future studies.

The beneficial association between antioxidant intake, bone mass, and the risk of osteoporosis has been studied in many countries, including the Republic of Korea [6,8,12,14,15], China [10,13], and Japan [9]. However, these studies primarily focused on the effects of single antioxidant intake, which might not fully explain the effects of all antioxidants in the diet. By calculating the TAC in the diet, this study evaluated the cumulative and overall antioxidant effects on bone mass and the risk of osteoporosis in the Korean female population.

To our knowledge, this is the first study to examine the association between dietary TAC and bone health in women with a relatively large sample size. The power verification had a value of more than 0.8, and the hypothesis verification of this study could be trusted. Rather than evaluating the effects of single antioxidants, the overall antioxidant capacity of food items was evaluated, and the accuracy of the assessments was improved. The limitations of this study include the following: (1) a cause-and-effect relationship between dietary TAC and the risk of osteoporosis could not be established due to the cross-sectional design of the original survey, (2) there is a possibility of potential residual confounders, and (3) there is a potential of assessment errors regarding food intake, since the one day of dietary records was based on participants' memory.

5. Conclusions

The present study showed that dietary TAC was positively associated with bone mass at the lumbar spine and femoral neck and inversely associated with the risk of osteoporosis in postmenopausal women. In premenopausal women, dietary TAC was positively associated with the BMC of the lumbar spine and total femur. These observations indicate that the consumption of foods with high dietary TAC, such as grapes, radish leaves, pepper paste, oranges, and spinach, could improve bone health, particularly in postmenopausal women. Additional long-term cohort or intervention studies are needed to confirm the effect of a diet with high dietary TAC on bone health and the risk of osteoporosis.

Supplementary Materials: The following are available online at https://www.mdpi.com/article/10.3390/nu13041149/s1, Supplement File S1: Dietary Total Antioxidant Capacity Calculator.

Author Contributions: D.K. performed data curation and formal analysis and prepared the original draft; A.H. performed validation and wrote the manuscript; Y.P. supervised the conceptualization, acquired funding, and reviewed and edited the manuscript. All authors have read and agreed to the published version of the manuscript.

Funding: This work was supported by the BK21 FOUR (Fostering Outstanding Universities for Research) project of the National Research Foundation of Korea Grant and a National Research Foundation of Korea (NRF) grant funded by the Korean government (MSIT) (NRF-2021R1A2B02002208).

Institutional Review Board Statement: The study was conducted according to the guidelines of the Declaration of Helsinki and approved by the Institutional Review Board of the Korea National Statistical Office, Centers for Disease Control and Prevention (KNHANES protocol), and the Hanyang University Institutional Review Board (No. HYUIRB-202101-007).

Informed Consent Statement: The need for patient consent was waived, as this study was conducted using Korea National Health and Nutrition Examination Survey 2008–2011 data that had already been investigated previously.

Data Availability Statement: Data from the Korea National Health and Nutrition Examination Survey (KNHANES) 2008–2011 can be found on the official website of the KNHANES (http://knhanes.cdc.go.kr, accessed on 30 March 2021).

Conflicts of Interest: The authors declare that there is no conflict of interest.

References

1. Klibanski, A.; Adams-Campbell, L.; Bassford, T.; Blair, S.N.; Boden, S.D.; Dickersin, K.; Gifford, D.R.; Glasse, L.; Goldring, S.R.; Hruska, K. Osteoporosis prevention, diagnosis, and therapy. *JAMA* **2001**, *285*, 785–795.
2. Park, E.J.; Joo, I.W.; Jang, M.J.; Kim, Y.T.; Oh, K.; Oh, H.J. Prevalence of osteoporosis in the Korean population based on Korea National Health and Nutrition Examination Survey (KNHANES), 2008–2011. *Yonsei Med. J.* **2014**, *55*, 1049–1057. [CrossRef]
3. Kim, Y.; Kim, J.H.; Cho, D.S. Gender difference in osteoporosis prevalence, awareness and treatment: Based on the Korea national health and nutrition examination survey 2008~2011. *J. Korean Acad. Nurs.* **2015**, *45*, 293–305. [CrossRef]
4. Hu, D.; Cheng, L.; Jiang, W. Fruit and vegetable consumption and the risk of postmenopausal osteoporosis: A meta-analysis of observational studies. *Food Funct.* **2018**, *9*, 2607–2616. [CrossRef]
5. Domazetovic, V.; Marcucci, G.; Iantomasi, T.; Brandi, M.L.; Vincenzini, M.T. Oxidative stress in bone remodeling: Role of antioxidants. *Clin. Cases Miner. Bone Metab.* **2017**, *14*, 209–216. [CrossRef]
6. Regu, G.M.; Kim, H.; Kim, Y.J.; Paek, J.E.; Lee, G.; Chang, N.; Kwon, O. Association between Dietary Carotenoid Intake and Bone Mineral Density in Korean Adults Aged 30–75 Years Using Data from the Fourth and Fifth Korean National Health and Nutrition Examination Surveys (2008–2011). *Nutrients* **2017**, *9*, 1025. [CrossRef]
7. Malmir, H.; Shab-Bidar, S.; Djafarian, K. Vitamin C intake in relation to bone mineral density and risk of hip fracture and osteoporosis: A systematic review and meta-analysis of observational studies. *Br. J. Nutr.* **2018**, *119*, 847–858. [CrossRef]
8. Kim, D.E.; Cho, S.H.; Park, H.M.; Chang, Y.K. Relationship between bone mineral density and dietary intake of beta-carotene, vitamin C, zinc and vegetables in postmenopausal Korean women: A cross-sectional study. *J. Int. Med. Res.* **2016**, *44*, 1103–1114. [CrossRef]
9. Odai, T.; Terauchi, M.; Hirose, A.; Kato, K.; Miyasaka, N. Bone Mineral Density in Premenopausal Women Is Associated with the Dietary Intake of alpha-Tocopherol: A Cross-Sectional Study. *Nutrients* **2019**, *11*, 2474. [CrossRef]
10. Zhang, Z.Q.; He, L.P.; Liu, Y.H.; Liu, J.; Su, Y.X.; Chen, Y.M. Association between dietary intake of flavonoid and bone mineral density in middle aged and elderly Chinese women and men. *Osteoporos. Int.* **2014**, *25*, 2417–2425. [CrossRef] [PubMed]
11. Hardcastle, A.C.; Aucott, L.; Reid, D.M.; Macdonald, H.M. Associations between dietary flavonoid intakes and bone health in a Scottish population. *J. Bone Miner. Res.* **2011**, *26*, 941–947. [CrossRef]
12. Kim, Y.A.; Kim, K.M.; Lim, S.; Choi, S.H.; Moon, J.H.; Kim, J.H.; Kim, S.W.; Jang, H.C.; Shin, C.S. Favorable effect of dietary vitamin C on bone mineral density in postmenopausal women (KNHANES IV, 2009): Discrepancies regarding skeletal sites, age, and vitamin D status. *Osteoporos. Int.* **2015**, *26*, 2329–2337. [CrossRef]
13. Zhang, J.; Zhang, K.; Shi, H.; Tang, Z. A cross-sectional study to evaluate the associations between hypertension and osteoporosis in Chinese postmenopausal women. *Int. J. Clin. Exp. Med.* **2015**, *8*, 21194.
14. Kim, M.H.; Lee, H.J. Osteoporosis, vitamin C intake, and physical activity in Korean adults aged 50 years and over. *J. Phys. Ther. Sci.* **2016**, *28*, 725–730. [CrossRef]
15. Park, H.M.; Heo, J.; Park, Y. Calcium from plant sources is beneficial to lowering the risk of osteoporosis in postmenopausal Korean women. *Nutr. Res.* **2011**, *31*, 27–32. [CrossRef]

16. Sugiura, M.; Nakamura, M.; Ogawa, K.; Ikoma, Y.; Yano, M. High Vitamin C Intake with High Serum beta-Cryptoxanthin Associated with Lower Risk for Osteoporosis in Post-Menopausal Japanese Female Subjects: Mikkabi Cohort Study. *J. Nutr. Sci. Vitaminol.* **2016**, *62*, 185–191. [CrossRef]
17. Wattanapenpaiboon, N.; Lukito, W.; Wahlqvist, M.L.; Strauss, B.J. Dietary carotenoid intake as a predictor of bone mineral density. *Asia Pac. J. Clin. Nutr.* **2003**, *12*, 12.
18. Floegel, A.; Kim, D.O.; Chung, S.J.; Song, W.O.; Fernandez, M.L.; Bruno, R.S.; Koo, S.I.; Chun, O.K. Development and validation of an algorithm to establish a total antioxidant capacity database of the US diet. *Int. J. Food Sci. Nutr.* **2010**, *61*, 600–623. [CrossRef]
19. Zujko, M.E.; Waskiewicz, A.; Witkowska, A.M.; Szczesniewska, D.; Zdrojewski, T.; Kozakiewicz, K.; Drygas, W. Dietary Total Antioxidant Capacity and Dietary Polyphenol Intake and Prevalence of Metabolic Syndrome in Polish Adults: A Nationwide Study. *Oxid. Med. Cell. Longev.* **2018**, *2018*, 7487816. [CrossRef]
20. Mozaffari, H.; Daneshzad, E.; Surkan, P.J.; Azadbakht, L. Dietary Total Antioxidant Capacity and Cardiovascular Disease Risk Factors: A Systematic Review of Observational Studies. *J. Am. Coll. Nutr.* **2018**, *37*, 533–545. [CrossRef]
21. Abbasalizad Farhangi, M.; Vajdi, M. Dietary Total Antioxidant Capacity (TAC) Significantly Reduces the Risk of Site-Specific Cancers: An Updated Systematic Review and Meta-Analysis. *Nutr. Cancer* **2020**, 1–19. [CrossRef]
22. Kweon, S.; Kim, Y.; Jang, M.J.; Kim, Y.; Kim, K.; Choi, S.; Chun, C.; Khang, Y.H.; Oh, K. Data resource profile: The Korea National Health and Nutrition Examination Survey (KNHANES). *Int. J. Epidemiol.* **2014**, *43*, 69–77. [CrossRef]
23. Rural Development Administration. *Food Composition Table*, 7th ed.; Rural Development Administration: Jeonju Deokjin, Korea, 2006. Available online: http://www.rda.go.kr/main/mainPage.do (accessed on 6 May 2020).
24. Lee, J.; Jang, S. A Study on Reference Values and Prevalence of Osteoporosis in Korea: The Korea National Health and Nutrition Examination Survey 2008–2011. *J. Korean Off. Stat.* **2013**, *18*, 42–65.
25. Park, S.-H.; Kim, S.-N.; Lee, S.H.; Choe, J.-S.; Choi, Y. Development of 9th Revision Korean Food Composition Table and Its Major Changes. *Korean J. Community Nutr.* **2018**, *23*. [CrossRef]
26. Kim, J.; Jang, H.; Choi, J. Flavonoid Data Base 1.0. Nongsaro. Volume 1. 2016. Available online: http://koreanfood.rda.go.kr/kfi/fct/fctCompSrch/list (accessed on 14 May 2020).
27. Bhagwat, S.; Haytowitz, D.; Wu, X. *USDA Database for the Flavonoid Content of Selected Foods, Release 3.3*, March 2018 ed.; USDA: Beltsville, MD, USA, 2018; Volume 3.3. Available online: http://www.ars.usda.gov/nutrientdata/flav (accessed on 14 May 2020).
28. Bhagwat, S.; Haytowitz, D.; Holden, J. USDA Database for the Isoflavone Content of Selected Foods, Release 2.1. 2008. Available online: https://www.ars.usda.gov/northeast-area/beltsville-md-bhnrc/beltsville-human-nutrition-research-center/methods-and-application-of-food-composition-laboratory/mafcl-site-pages/isoflavone/ (accessed on 14 May 2020).
29. Bhagwat, S.; Haytowitz, D.; Holden, J. USDA Database for the Proanthocyanidin Content of Selected Foods, Release 2.1. 2018. Available online: https://www.ars.usda.gov/northeast-area/beltsville-md-bhnrc/beltsville-human-nutrition-research-center/methods-and-application-of-food-composition-laboratory/mafcl-site-pages/proanthocyanidin/ (accessed on 14 May 2020).
30. Ryu, S.N.; Han, S.J.; Park, S.Z.; Kim, H.Y. Antioxidative activity and varietal difference of cyanidin 3-glucoside and peonidin 3-glucoside contents in pigmented rice. *Korean J. Crop. Sci.* **2000**, *45*, 257–260.
31. Schauss, A.G.; Wu, X.; Prior, R.L.; Ou, B.; Patel, D.; Huang, D.; Kababick, J.P. Phytochemical and nutrient composition of the freeze-dried Amazonian palm berry, Euterpe oleraceae Mart.(Acai). *J. Agric. Food Chem.* **2006**, *54*, 8598–8603. [CrossRef]
32. Sakakibara, H.; Honda, Y.; Nakagawa, S.; Ashida, H.; Kanazawa, K. Simultaneous determination of all polyphenols in vegetables, fruits, and teas. *J. Agric. Food Chem.* **2003**, *51*, 571–581. [CrossRef]
33. Wen, P.; Hu, T.-G.; Linhardt, R.J.; Liao, S.-T.; Wu, H.; Zou, Y.-X. Mulberry: A review of bioactive compounds and advanced processing technology. *Trends Food Sci. Technol.* **2019**, *83*, 138–158. [CrossRef]
34. Jun, S.; Chun, O.K.; Joung, H. Estimation of dietary total antioxidant capacity of Korean adults. *Eur. J. Nutr.* **2018**, *57*, 1615–1625. [CrossRef]
35. Saylor, J.; Friedmann, E.; Lee, H.J. Navigating complex sample analysis using national survey data. *Nurs. Res.* **2012**, *61*, 231–237. [CrossRef]
36. Greenland, S.; Pearce, N. Statistical foundations for model-based adjustments. *Annu. Rev. Public Health* **2015**, *36*, 89–108. [CrossRef]
37. Li, J.J.; Huang, Z.W.; Wang, R.Q.; Ma, X.M.; Zhang, Z.Q.; Liu, Z.; Chen, Y.M.; Su, Y.X. Fruit and vegetable intake and bone mass in Chinese adolescents, young and postmenopausal women. *Public Health Nutr.* **2013**, *16*, 78–86. [CrossRef]
38. Zeng, L.F.; Luo, M.H.; Liang, G.H.; Yang, W.Y.; Xiao, X.; Wei, X.; Yu, J.; Guo, D.; Chen, H.Y.; Pan, J.K.; et al. Can Dietary Intake of Vitamin C-Oriented Foods Reduce the Risk of Osteoporosis, Fracture, and BMD Loss? Systematic Review With Meta-Analyses of Recent Studies. *Front. Endocrinol.* **2019**, *10*, 844. [CrossRef]
39. Rozman, C.; Feliu, E.; Berga, L.; Reverter, J.; Climent, C.; Ferran, M. Age-related variations of fat tissue fraction in normal human bone marrow depend both on size and number of adipocytes: A stereological study. *Exp. Hematol.* **1989**, *17*, 34–37.
40. Li, J.; Chen, X.; Lu, L.; Yu, X. The relationship between bone marrow adipose tissue and bone metabolism in postmenopausal osteoporosis. *Cytokine Growth Factor Rev.* **2020**, *52*, 88–98. [CrossRef]
41. Wright, J.A. A comparison of rat femoral, sternebral and lumbar vertebral bone marrow fat content by subjective assessment and image analysis of histological sections. *J. Comp. Pathol.* **1989**, *100*, 419–426. [CrossRef]

42. Macdonald, H.M.; New, S.A.; Golden, M.H.; Campbell, M.K.; Reid, D.M. Nutritional associations with bone loss during the menopausal transition: Evidence of a beneficial effect of calcium, alcohol, and fruit and vegetable nutrients and of a detrimental effect of fatty acids. *Am. J. Clin. Nutr.* **2004**, *79*, 155–165. [CrossRef]
43. Osterhoff, G.; Morgan, E.F.; Shefelbine, S.J.; Karim, L.; McNamara, L.M.; Augat, P. Bone mechanical properties and changes with osteoporosis. *Injury* **2016**, *47*, S11–S20. [CrossRef]
44. Berriche, O.; Chiraz, A.; Othman, R.B.; Souheila, H.; Lahmer, I.; Wafa, C.; Sebai, I.; Sfar, H.; Mahjoub, F.; Jamoussi, H. Nutritional risk factors for postmenopausal osteoporosis. *Alex. J. Med.* **2019**, *53*, 187–192. [CrossRef]
45. Jones, S.R.; Carley, S.; Harrison, M. An introduction to power and sample size estimation. *Emerg. Med. J.* **2003**, *20*, 453–458. [CrossRef]
46. Schiessl, H.; Frost, H.M.; Jee, W.S.S. Estrogen and Bone-Muscle Strength and Mass Relationships. *Bone* **1998**, *22*, 1–6. [CrossRef]
47. Nakano, M.; Nakamura, Y.; Suzuki, T.; Miyazaki, A.; Takahashi, J.; Saito, M.; Shiraki, M. Pentosidine and carboxymethyl-lysine associate differently with prevalent osteoporotic vertebral fracture and various bone markers. *Sci. Rep.* **2020**, *10*, 22090. [CrossRef]
48. Behera, J.; Bala, J.; Nuru, M.; Tyagi, S.C.; Tyagi, N. Homocysteine as a pathological biomarker for bone disease. *J. Cell. Physiol.* **2017**, *232*, 2704–2709. [CrossRef]
49. Iguacel, I.; Miguel-Berges, M.L.; Gómez-Bruton, A.; Moreno, L.A.; Julián, C. Veganism, vegetarianism, bone mineral density, and fracture risk: A systematic review and meta-analysis. *Nutr. Rev.* **2019**, *77*, 1–18. [CrossRef] [PubMed]
50. Melina, V.; Craig, W.; Levin, S. Position of the Academy of Nutrition and Dietetics: Vegetarian diets. *J. Acad. Nutr. Diet.* **2016**, *116*, 1970–1980. [CrossRef] [PubMed]

Review

Botanicals in Postmenopausal Osteoporosis

Wojciech Słupski, Paulina Jawień and Beata Nowak *

Department of Pharmacology, Wroclaw Medical University, ul. J. Mikulicza-Radeckiego 2,
50-345 Wrocław, Poland; wojciech.slupski@umed.wroc.pl (W.S.); paulina.jawien@umed.wroc.pl (P.J.)
* Correspondence: beata.nowak@umed.wroc.pl; Tel.: +48-607-924-471

Abstract: Osteoporosis is a systemic bone disease characterized by reduced bone mass and the deterioration of bone microarchitecture leading to bone fragility and an increased risk of fractures. Conventional anti-osteoporotic pharmaceutics are effective in the treatment and prophylaxis of osteoporosis, however they are associated with various side effects that push many women into seeking botanicals as an alternative therapy. Traditional folk medicine is a rich source of bioactive compounds waiting for discovery and investigation that might be used in those patients, and therefore botanicals have recently received increasing attention. The aim of this review of literature is to present the comprehensive information about plant-derived compounds that might be used to maintain bone health in perimenopausal and postmenopausal females.

Keywords: osteoporosis; menopause; botanicals; herbs

Citation: Słupski, W.; Jawień, P.; Nowak, B. Botanicals in Postmenopausal Osteoporosis. *Nutrients* **2021**, *13*, 1609. https://doi.org/10.3390/nu13051609

Academic Editor: Jose M. Moran

Received: 9 April 2021
Accepted: 10 May 2021
Published: 11 May 2021

Publisher's Note: MDPI stays neutral with regard to jurisdictional claims in published maps and institutional affiliations.

1. Introduction

Women's health and quality of life is modulated and affected strongly by hormone status. An oestrogen level that changes dramatically throughout life determines the development of women's age-associated diseases. Age-associated hormonal imbalance and oestrogen deficiency are involved in the pathogenesis of various diseases, e.g., obesity, autoimmune disease and osteoporosis. Many female patients look for natural biological products deeply rooted in folk medicine as an alternative to conventional pharmaceutics used as the prophylaxis of perimenopausal health disturbances. This review will focus on botanicals and plant derived substances that may be used to maintain bone health in perimenopausal and postmenopausal females.

Osteoporosis is a systemic bone disease characterized by the reduced bone mass and deterioration of bone microarchitecture leading to bone fragility and the increased risk of fractures [1]. Osteoporosis-associated fragility fractures constitute a major health problem all over the world. It is estimated that more than 40 million American citizens over 50 years of age are at risk of osteoporotic fractures, and that due to the demographic changes, this number will at least double until the year 2040 [2]. It is also predicted that 25% of people over 50 who have experienced osteoporotic hip fracture will die within a year [2]. Hypogonadism, mainly associated with menopause, is the main cause of osteoporosis. High social and individual costs of osteoporosis and its complications remain a challenge for health systems, especially because most of the patients with osteoporosis remain untreated. The data indicate that almost 60% patients at high risk of osteoporotic fractures are not receiving osteoprotective treatment [3]. Additionally, a decrease in the usage of antiosteoporotic drugs, especially bisphosphonates, has been observed in recent years [3]. Oral bisphosphonates, that bind to hydroxyapatite and inhibit osteoclastic bone resorption, are the drug of choice for the treatment of primary osteoporosis. However, they are associated with side effects including oesophagitis and oesophageal ulcers, jaw osteonecrosis, and atypical femoral fractions. In case of intolerance or lack of efficacy, they might be switched to intravenous bisphosphonates, strontium ranelate, denosumab, teriparatide, abaloparatide or romosozumab. As additional options in postmenopausal women, raloxifene and hormonal replacement therapy may be used [4]. However, as those pharmaceutics are associated with various side effects, many women seek for botanicals as an alternative therapy.

Bones undergo continuous remodelling, osteoblasts synthetize the bone matrix and, at the same time, osteoclasts degrade bone tissue. In physiological conditions, we observe the balance between the resorption and formation of bone tissue. This balance depends on the activity, differentiation, and apoptosis of bone forming osteoblasts and bone-resorbing osteoclasts. Multiple factors and signalling pathways modulate bone homeostasis (Figure 1). Bone cells' activity is controlled, among others, by growth factors (IGF—insulin-like growth factor, TGFβ—tumour growth factor β, PDGF—platelet-derived growth factor), bone morphogenic proteins (BMPs), hormones (parathormone, thyroid hormones, sex hormones, insulin, prolactin, growth hormone) and vitamins (vitamin D). Wnt, BMPs and TGFβ pathways interact with other signalling molecules such as basic fibroblast growth factor (bFGF), Hedgehog (Hh) and IGF-1, and regulate the differentiation and activity of osteoclasts [5]. Runx2 (Runt-related transcription factor 2) and OSX (Osterix) are the main transcription factors involved in the modulation of osteoblast differentiation. Osteoclastogenesis is regulated by two main pathways: RANK/RANKL (Receptor Activator for Nuclear Factor κB/Receptor Activator for Nuclear Factor κB Ligand) and M-CSF/c-FMS (the macrophage colony-stimulating factor/colony-stimulating factor-1 receptor) system. Parathyroid hormone (PTH) and calcitriol induce RANKL synthesis in osteoblasts and afterwards promote osteoclastogenesis through RANK activation. RANK activation is counteracted by OPG (osteoprotegerin), which is a decoy receptor of free RANKL. M-CSF/c-FMS interaction leads to mitogen-activated protein kinase (MAPK) activation that induces RANKL production and activates AKT/mTOR (protein kinase B/mechanistic target of rapamycin) pathway engaged in the survival of osteoclasts [5].

Oestrogen plays an important role in maintaining bone mineral density in both rodents and humans (Figure 2). A decrease in the oestrogen level associated with menopause leads to a decrease in bone mineral density (BMD) that increases the risk of fractures [6,7]. The protective effect of oestrogen in bone is due to many mechanisms. Oestrogen, among other things, inhibits bone resorption by the suppression of the synthesis of proinflammatory cytokines in osteoblasts via the inhibition of nuclear factor-kappa B (NFκB) signalling pathway [8]. They also activate the transcription of a gene encoding Fas Ligand (FasL) in osteoblasts. Soluble FasL (sFasL) released from the osteoblast binds to the transmembrane Fas receptor (FasR) on the osteoclast's surface and induces the apoptosis of osteoclasts [9]. Additionally, oestrogen decreases the RANKL/OPG ratio and prevents bone resorption [10].

Women's health and quality of life are modulated and affected strongly by hormone status. An oestrogen level that changes dramatically determines the development of women's age-associated diseases. Age-associated hormonal imbalance and oestrogen deficiency are involved in the pathogenesis of various diseases, e.g., obesity, autoimmune diseases, and osteoporosis. As postmenopausal osteoporosis is characterised by bone resorption that exceeds bone formation, antiresorptive drugs are one of the therapeutic options and most current therapies exert mainly antiresorptive effects. Another therapeutic solution may be the use of anabolic drugs that would enhance bone formation. Bone morphogenic protein (BMP), Wnt, and insulin-like growth factor 1 (IGF1) are the key molecules involved in the regulation of osteoblast formation and activation [11–13]. Oestrogens, SERMs (selective oestrogen receptor modulators), bisphosphonates, strontium ranelate, denosumab, teriparatide, abaloparatide or romosozumab are clinically used as effective therapies against postmenopausal osteoporosis [4]; however, their usage is associated with the established risk of the side effect. Therefore, many female patients look for natural biological products deeply rooted in folk medicine as an alternative to conventional pharmaceutics used as the prophylaxis of perimenopausal health disturbances. This review will focus on botanicals and plant-derived substances that may be used to maintain bone health in perimenopausal and postmenopausal females. The aim of the review is to present the currently available results of clinical and preclinical studies, investigating the influence of plant-derived extracts and compounds on menopause-associated disturbances of bone metabolism. For the purpose of the article, we defined botanicals as substances obtained or derived from plants, such as a plant part or the extract, or compounds isolated from plants or their extracts. While searching for the information in PubMed and Google Scholar, we tried not to limit our research to Chinese traditional medicine, but to broaden it by including less known European plants, e.g., *Humulus lupulus* L. or *Equisetum arvense* L. We focused on the research reported after 2010, but we did not exclude earlier studies in our review. Table 1 summarizes the information about the main active ingredients discussed in the article, and Table 2 clinical studies and their main findings.

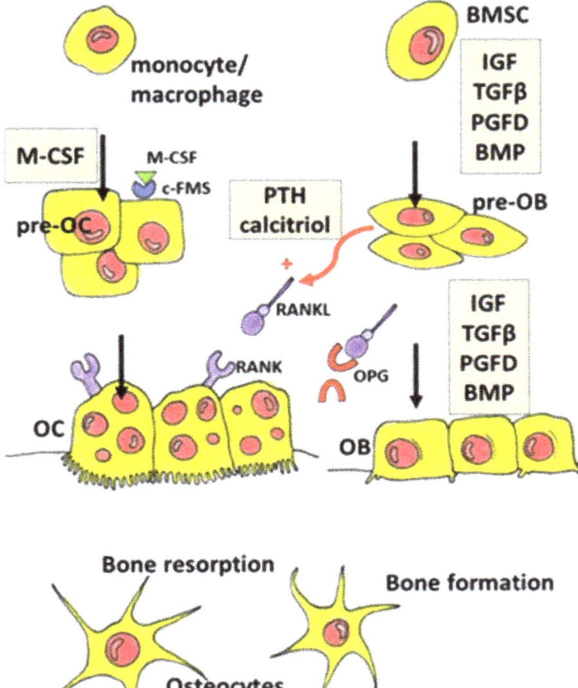

Figure 1. Schematic diagram representing regulation of osteoblast and osteoclast differentiation. BMP—bone morphogenic protein, BMSC—bone marrow-derived mesenchymal stem cells, c-FMS—colony-stimulating factor-1 receptor, IGF—insulin-like growth factor, M-CSF—macrophage colony-stimulating factor, OB—osteoblast, OC—osteoclast, OPG—osteoprotegerin, PGFD—platelet-derived growth factor, pre-OB—pre-osteoblasts, pre-OC—pre-osteoclasts, PTH—parathyroid hormone, RANK—Receptor Activator for Nuclear Factor κB, RANKL—Receptor Activator for Nuclear Factor κB Ligand, TGFβ—tumour growth factor β.

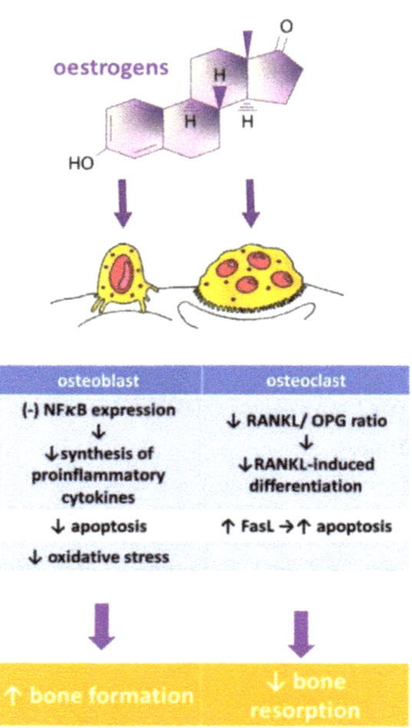

Figure 2. Influence of oestrogens on osteoblast and osteoclast function, and bone turnover. FasL—Fas Ligand, NFκB—Nuclear Factor κB, OPG—osteoprotegerin, RANKL—Receptor Activator for Nuclear Factor κB Ligand, ↑—increased, ↓—decreased

Table 1. Herbal compounds with antiosteoporotic properties investigated in vitro and in animal models.

Herbal Compounds	Subgroup	Chemical Structure	Proposed Mechanism of Action
Daidzein	isoflavones		ER mediated signalling pathway, activation of intracellular pathways: AKT, phospholipase C (PLC), mitogen-activated protein kinase (MAPK) [14]
Genistein	isoflavones		ER-mediated signalling pathway, activation of intracellular pathways: AKT, PLC, MAPK [14]
Ipriflavone	isoflavones		Modulation of key signalling pathways to regulate bone resorption (e.g., ↓urinary DPD, NTX) and bone formation (e.g., ↑BALP and osteocalcin [15]

Table 1. *Cont.*

Herbal Compounds	Subgroup	Chemical Structure	Proposed Mechanism of Action
Biochanin A	O-methylated isoflavones	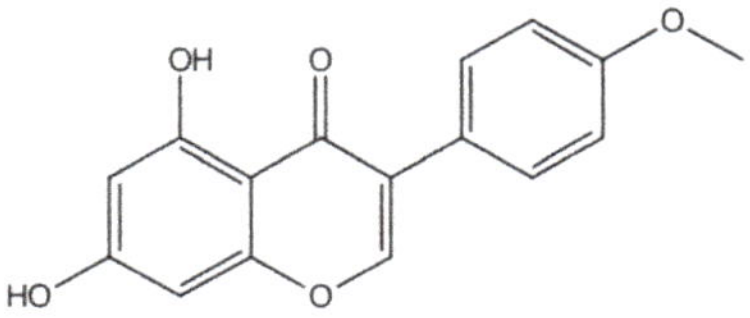	ER mediated signalling pathway, activation of intracellular pathways: AKT, PLC, MAPK [14]
Formononetin	O-methylated isoflavones	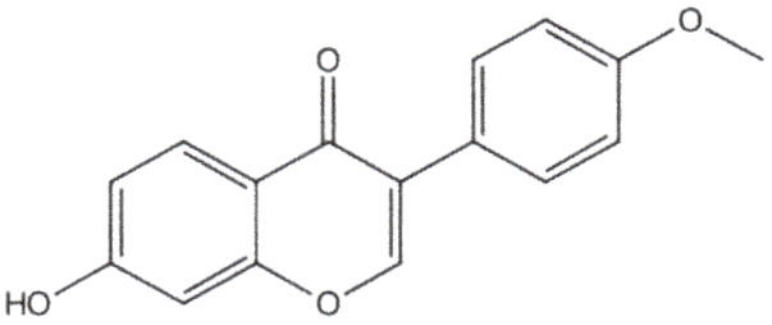	ER mediated signalling pathway, activation of intracellular pathways: AKT, PLC, MAPK [14]
Glycitein	O-methylated isoflavones	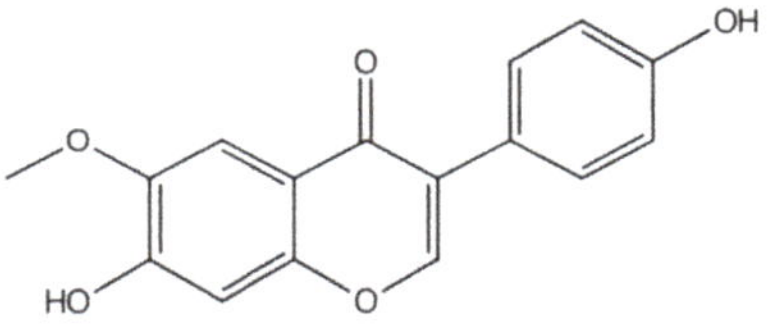	ER mediated signalling pathway, activation of intracellular pathways: AKT, PLC, MAPK [14]
Icariin	prenylated flavonol glycoside	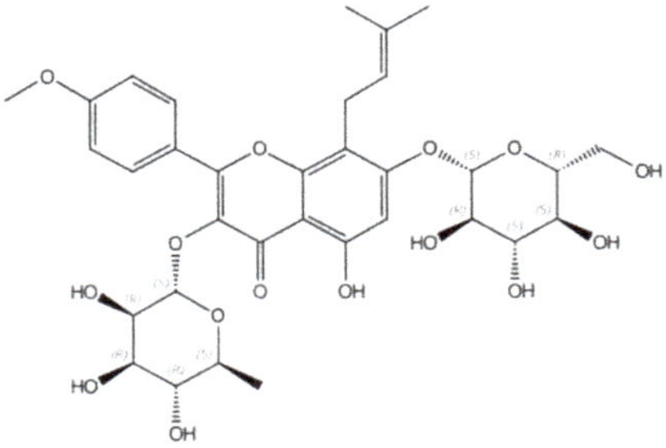	Stimulation of bone formation by promotion of osteoblasts differentiation and enhancement of their activity [16]; activation of BMP-2/Smad4, Wnt and IGF-1 signal transduction pathways [5,17], induction of ERK, JNK and p38 kinase activation [18]; decreasing of RANKL-induced osteoclastogenesis via inhibition of NFκB and MAPK expression [19]
8-prenylnaringenin	prenylflavonoids	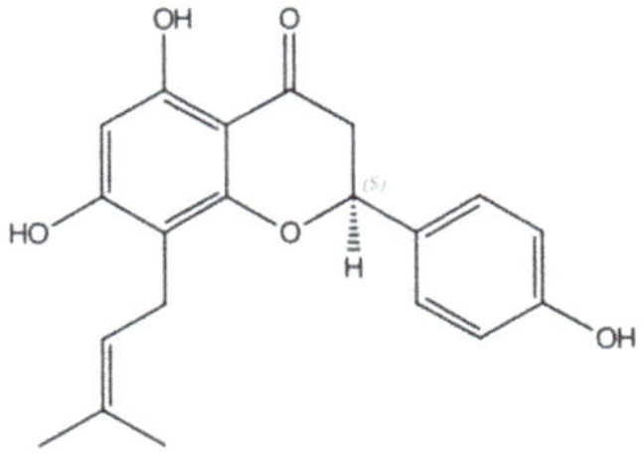	Promotion of osteoblast differentiation and induction of osteoclast apoptosis [20]
Epimedin B	prenylflavonoids	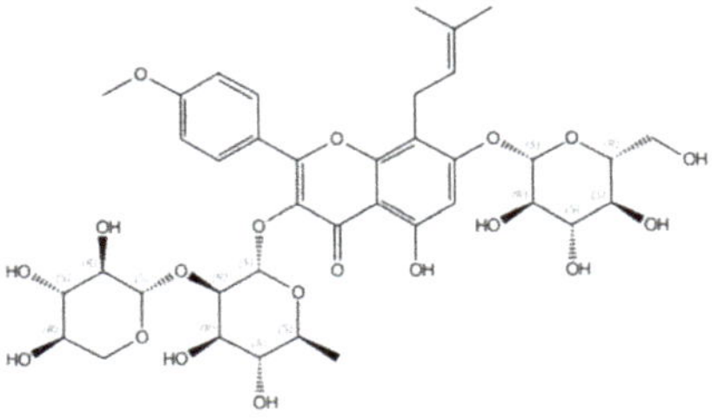	Inhibition of bone resorption, bone formation promotion and urinary calcium excretion blocking [21]

Table 1. *Cont.*

Herbal Compounds	Subgroup	Chemical Structure	Proposed Mechanism of Action
Epimedin C	prenylflavonoids	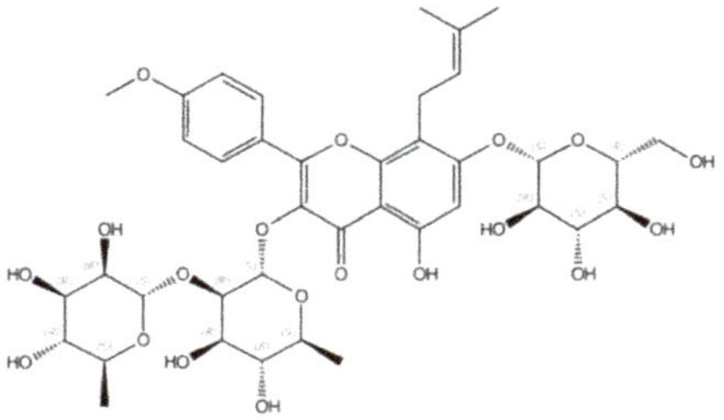	Inhibition of bone resorption, bone formation promotion and urinary calcium excretion blocking [21]
Tanshinones (dihydrotanshinone, tanshinone I, or tanshinone IIA)	diterpenes	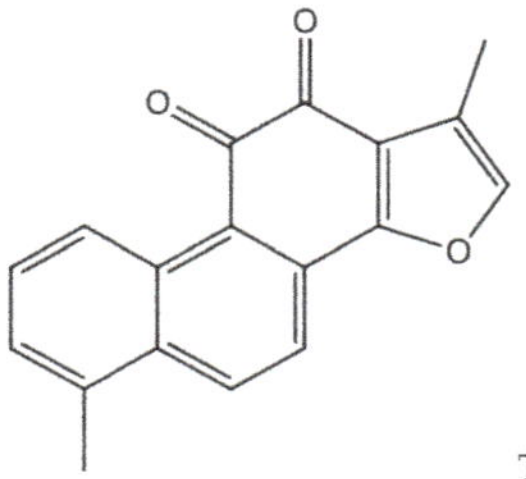 Tanshinon 1	Inhibition of the TRAP5b-expressing osteoclasts formation by suppressing RANKL-induced expression of c-fos and NFATc1 [22,23]
Salvianolic acid A	phenolic acids	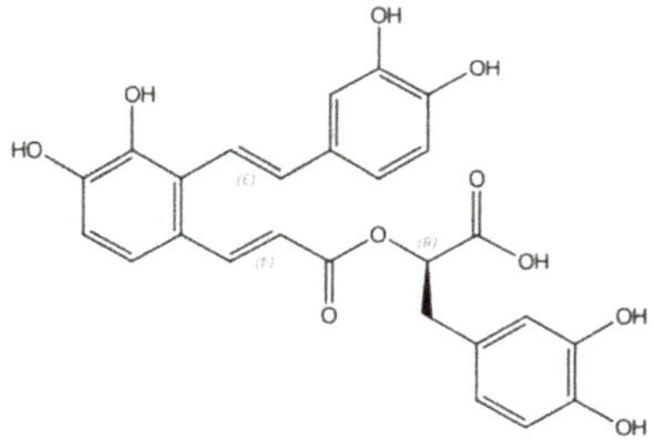	osteoblast differentiation modulation and osteoblast activity upregulation [24,25]
Salvianolic acid B	phenolic acids	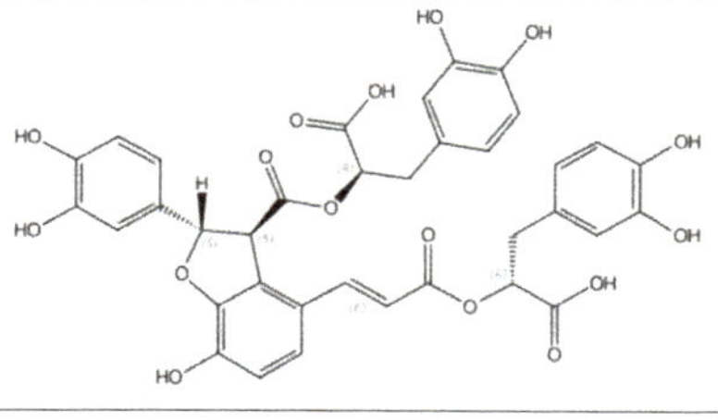	osteoblast differentiation modulation and osteoblast activity upregulation [24,25]
Eudebeiolide B	eudesmane-type sesquiter-penoid	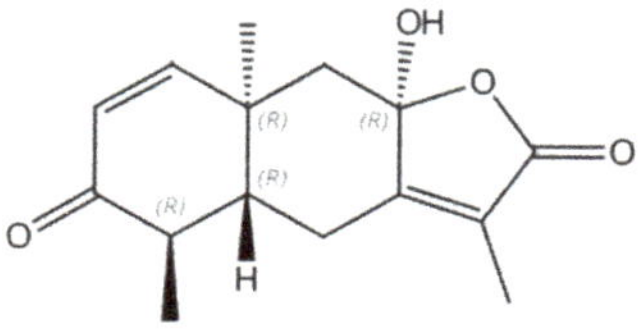	Osteoclastogenesis inhibition and ovariectomy-induced bone loss prevention by regulating RANKL-Induced NF-κB, c-Fos and Calcium Signaling [26]

AKT—protein kinase B, BALP—bone-specific alkaline phosphatase, CTX—type I collagen cross-linked C-telopeptide, DPD—deoxypyridinoline, ER—oestrogen receptor, ERK—extracellular signal-regulated kinase, JNK—c-Jun N-terminal kinase, MAPK—mitogen-activated protein kinase, NFκB—nuclear factor-kappa B, NTX—type I collagen cross-linked N-telopeptide, PLC—phospholipase C, RANKL—Receptor Activator for Nuclear Factor κB Ligand, TRAP 5b—Tartrate-resistant acid phosphatase 5b.

Table 2. Summary of potential anti-osteoporotic properties of botanicals in clinical trials.

Botanicals	Population and Design	Intervention	Outcome	Authors and References
Soy isoflavones	single open-group prospective clinical intervention; 42 postmenopausal women,	three daily servings for 12 consecutive weeks of whole soy foods containing approximately 60 mg/day of isoflavones	↓ NTX, ↑ osteocalcin	Scheiber 2001 [27]
Soy isoflavones	RCT with 3 groups: soy rich diet, HRT, control; 187 healthy asymptomatic postmenopausal women aged 39–60,	approximately 47 mg/day of isoflavones in diet group; duration: 6 moths	↑ bone osteoblastic activity but not as effective as HRT in reducing the postmenopausal turnover, ↑ osteocalcin	Chiechi 2002 [28]
Soy isoflavones	RCT with 3 groups: placebo, mid-dose, and high-dose, in pill form; 203 postmenopausal Chinese women aged 48 to 62,	placebo (daily dose of 0 mg isoflavones + 500 mg calcium, $n = 67$) mid-dose (40 mg isoflavones + 500 mg calcium, $n = 68$) and high-dose (80 mg isoflavones + 500 mg calcium, $n = 68$); duration: 12months	favourable effect on rates of change in BMC at the total hip and trochanter among later postmenopausal women (>4 y), in women with lower body weight (≤median, 55.5 kg), or among women with lower level of calcium intake (≤median, 1095 mg/day)	Chen 2004 [29]
Soy isoflavones	RCT with 3 groups: placebo, mid-dose, and high-dose; 90 Chinese postmenopausal women aged 45–60	placebo (daily dose of 0 mg isoflavones) mid-dose (84 mg) and high dose (126 mg), 30 subjects/group; duration: 6months	Retardation of lumbar and femoral bone loss at the lumbar spine (L1–L4) and bone resorption	Ye 2006 [30]
Soy isoflavones	double-blind RCT with 2 groups: placebo, isoflavone conjugates in capsule form, 68 postmenopausal Japanese women	Isoflavone group (75 mg of isoflavone conjugates/day), 34 subjects/group; duration: 12 months	↑ serum equol in the equol producers but not in the nonproducers, preventive effects of isoflavones on hip BMD	Wu 2007 [31]
Soy isoflavones	double-blind RCT with 3 groups: placebo, mid-dose, and high-dose in tablet form; 255 postmenopausal women aged 46–63	placebo (daily dose of 0 mg isoflavones) mid-dose (80 mg) and high dose (120 mg); duration: 3 years	mild beneficial femoral BMD—and SSI	Shedd-Wise 2011 [32]
Soy isoflavones	double-blind RCT with 2 groups: placebo, isoflavones in tablet form; 87 Korean postmenopausal women aged 45–60	Isoflavone group = 70 mg in 2 tablet per day (8.0 mg glycitin, 20 mg daidzein, and 12.4 mg genistin); duration: 12 weeks	↑ serum BALP and osteocalcin	Lee 2017 [33]
Soy isoflavones	RCT with 3 groups; placebo, HRT, phytoestrogens; 325 postmenopausal women	HRT group (1 mg oestradiol and 0.5 mg norethisterone acetate p.o. daily, phytoestrogens group (40% standardized extract with 20 mg soy isoflavones (genistein and daidzein), two capsules = 40 mg p.o. daily; duration: 12 months	no significant differences between the effectiveness of the HRT and phytoestrogen in terms of effects on BMD and bone resorption	Tit 2018 [34]
Soy isoflavones	double-blind RCT with 3 groups: placebo, soy protein, soy protein + isoflavone in snack bar; 200 women within 2 years of the onset of their menopause	placebo (isoflavone of less than 300 parts per billion) PI (15 g soy protein with 66 mg of isoflavones), SP (15 g soy protein alone, isoflavone free) daily, 100 women/group; duration: 6 months	↓CTX with SPI supplementation compared to SP, ↓ P1NP with SPI supplementation	Sathyapalan 2017 [35]
Soy isoflavones	double-blind RCT with 2 groups: placebo, isoflavones in form of tablet	placebo (0 mg of isoflavones), isoflavones extracted from soy protein (200 mg daily = 4tablets) 248 multi-ethnic menopausal women aged 45 to 60; duration: 2 years	not superior to placebo in preventing bone loss or in reducing bone turnover or menopausal symptoms in women in the first 5 years of menopause	Levis 2011 [36]
Soy isoflavones	double-blind RCT with 2 groups: placebo, phytoestrogens; 202 postmenopausal women aged 60–75	placebo (milk protein), phytoestrogens (25.6 g soy protein containing 52 mg genistein, 41 mg daidzein and 6 mg glycetein (aglycone weights; duration: 12 months	no significant differences for BALP, calcium, and phosphorus measurements.	Kreijkamp-Kaspers 2004 [37]

Table 2. *Cont.*

Botanicals	Population and Design	Intervention	Outcome	Authors and References
Soy isoflavones	double-blind, multicentre RCT with 2 groups: isoflavone-enriched biscuits and bars and control biscuits and bars; 237 early postmenopausal women aged 53 ± 3y	placebo group (biscuits and cereal bar), isoflavone- enriched foods (soy isoflavone concentrate containing 40–50% of isoflavones) providing a mean daily intake of 110 mg isoflavone aglycones/day; duration: 12 months	isoflavone-enriched products did not alter lumbar and total body BMD or markers of bone formation and bone resorption	Brink 2008 [38]
Genistein	double-blind RCT with 2 groups: placebo, genistein; 389 postmenopausal women	placebo group (calcium and vitamin D, $n = 191$), genistein aglycone group (54 mg/day + calcium and vitamin D, $n = 198$) duration: 36 months	↑lumbar and femoral BMD, ↓bone resorption markers (DPD, CTX, RANKL), ↑ bone formation markers (BALP, IGF-1 and OPG)	Marini 2007 [39]; Marini 2008 [40]
Genistein	double-blind RCT with 2 groups: placebo, genistein; 138 postmenopausal women (age 49–67 years)	placebo (0mg of isoflavones, $n = 67$), genistein (54 mg/day, $n = 71$), duration: 24 months	↑ femoral and lumbar BMD, improvement of the quantitative ultrasound parameters (stiffness index, amplitude-dependent speed of sound, and bone transmission time)	Atteritano 2009 [41]
Genistein	double-blind RCT with 2 groups: placebo, geniVida™ bone blend group; 70 postmenopausal women	placebo (calcium only, $n = 28$), genistein group = 30 mg/daygenistein + vitamin D3 (800 IU/days) + vitamin K1 (150 µg/days) + polyunsaturated fatty acids (1 g polyunsaturated fatty acids as ethyl ester: eicosapentaenoic acid/docosahexaenoic acid ratio = ~2/1, $n = 30$); duration: 6 months	↑ BMD, ↑ BALP and NTX	Lappe 2013 [42]
Genistein	double-blind RCT with 2 groups: placebo, genistein, 121 postmenopausal women	placebo (1000 mg of calcium and 800 IU vitamin D3; $n = 59$), genistein aglycone group (54 mg/day + calcium, vitamin D3; $n = 62$, duration: 24 months	↑ femoral and lumbar BMD, ↑ BALP	Arcoraci 2017 [43]
Red clover isoflavones (genistein, daidzein, formononetin, biochanin A)	double-blind RCT with 4 groups: placebo, red clover isoflavone preparation (Rimostil) in 3 doses, 46 postmenopausal women	placebo, Rimostil (phytoestrogens)—28.5 mg, 57 mg, or 85.5mg/day, duration: 6 months,	↑ BMD after 57 mg and 85.5 mg/day	Clifton-Bligh 2001 [44]
Red clover isoflavones	double-blind RCT with 2 groups: placebo, isoflavone supplement Promensil®; 205 pre-, peri-, and postmenopausal women aged 49–65	placebo, isoflavone supplement (providing 26 mg biochanin A, 16 mg formononetin, 1 mg genistein, 0.5 mg daidzein/daily); duration: 12 months	↑ bone formation markers (BALP, P1NP), ↓ lumbar spine BMC and BMD	Akinson 2004 [45]
Red clover isoflavones	double-blind, parallel RCT with 2 groups: placebo, red clover extract; 78 postmenopausal osteopenic women supplemented with calcium 1200 mg/day, magnesium 550 mg/day, calcitriol 25 µg/day	placebo, red clover extract (60 mg isoflavone aglycones/day + probiotics); duration: 12 months	↓ lumbar and femoral BMD loss, ↓ CTX	Lambert 2017 [46]
Red clover isoflavones	double-blind RCT with 2 groups: placebo, red clover extract; 60 menopausal women	placebo, red clover extract (daily dose of 150 mL containing 37.1 mg isoflavones = 33.8 mg as aglycones); duration: 12 weeks	↑ spinal BMD	Thorup 2015 [47]
Red clover isoflavones	double-blind RCT with 2 groups: placebo, standardized red clover isoflavone dietary supplement (Promensil®); 401 healthy women aged 35–70 years	Placebo, red clover isoflavones (40 mg/day); duration: 36 months	safe and well tolerated but no effect on BMD	Powles 2008 [48]

Table 2. *Cont.*

Botanicals	Population and Design	Intervention	Outcome	Authors and References
Red clover isoflavones	double-blind RCT with 3 groups: placebo and 2 dietary supplements derived from red clover, 252 menopausal women ages 45–60 years	placebo, Promensil® (82 mg total isoflavones), Rimostil® (57.2 mg total isoflavones), duration: 12 weeks	no effect on bone turnover markers.	Knudson Schult 2004 [49]
Kudzu root (*Pueraria candollei var. mirifica*)	double-blind RCT with 4 groups: placebo, 3 dose of Pueraria; 71 postmenopausal women aged 45 to 60 years	placebo (*n* = 20), Pueraria mirifica in capsules (20, 30, or 50 mg once daily, *n* = 51); duration: 24 weeks	↓ BALP	Manonai 2008 [50]
Kudzu root (*Pueraria candollei* var. *mirifica*)	double-blind RCT with 2 groups 19 postmenopausal women	placebo tablet, tablet containing 25 mg dried PM root powder, 4 tablets/day; duration: 2 months	↓ ALP	Okamura 2008 [51]
Epimedium	double-blind RCT with 2 groups: placebo, Epimedium-derived phytoestrogen flavonoids (EPF), 100 healthy late postmenopausal women	placebo (*n* = 50), EPF group (*n* = 50; a daily dose of 60 mg Icariin, 15 mg daidzein, and 3 mg genistein), +300 mg calcium daily for both group; duration: 24 months	↑ lumbar and femoral BMD, ↓ DPD,	Zang 2007 [52]
Dried plums	RCT with 2 groups: placebo (dried apples), dried plums; 58 postmenopausal women	placebo (dried apples 75 g daily), dried plums (100 g daily); duration: 3 months	↑IGF-1, ↑ ALP, ↑ BALP	Ajamandi 2002 [53]
Dried plums	RCT with 2 groups: placebo, dried plums, 160 postmenopausal women with osteopenia	placebo (dried apples 75 g daily), dried plums (100 g daily) + 500 mg Calcium, 400 IU (10 μg) vitamin D daily for both group; duration: 12 months	↑ ulnar and lumbar BMD, ↓ BALP	Hooshmand 2011 [54]
Dried plums	RCT with 3 groups: placebo, 2 dose of dried plums, 48 older postmenopausal women	control (0 g/day dried plum), dried plum (50 or 100 g/day dried plum), duration: 6 months	↑ BMD, ↓ TRAP-5b, ↑ BALP/TRAP-5b ratio	Hooshmand 2016 [55]
Dried plums	RCT with 3 groups: placebo, 2 dose of dried plums; 35 men between the ages of 55 and 80 with moderate bone loss	control group (0g prunes), 100 g prunes daily, 50 g prunes daily, + multivitamin containing 450 mg calcium and 800 IU vitamin D for all group, duration: 3 months	↓ osteocalcin, ↑ OPG/RANKL ratio	Ajmandi 2020 [56]
Horsetail (*Equisetum arvense*)	Double blind RCT with 4 groups: control, placebo + horsetail extract, horsetail extract, calcium, 122 women in menopause for at least two years	no treatment/control group (*n* = 29), placebo for 40 days and titrated horsetail extract for a further 40 days (*n* = 31), titrated dry horsetail extract for 80 days (*n* = 30); Calcium (Osteosil®) for 80 days (*n* = 32), After the 80-day initial study period, patients treated with titrated horsetail extract and with calcium continued treatment for one year	↑ in the average densitometric values for the vertebra	Corletto 1999 [57]
Black cohosh (*Cimcifuga racemosa*)	double-blind RCT with 3 groups: placebo, black cohosh, oestrogens; 62 postmenopausal women	placebo, black cohosh (40 mg of herbal drug/day), conjugated oestrogens (0.6 mg/day); duration: 12 weeks.	↑ osteoblast activity, weak estrogen-like activity, no significant effects on coagulation markers and liver enzymes	Wuttke 2006 [58]
Black cohosh (*Cimcifuga racemosa*)	prospective clinical trial with 2 groups: untreated control, isopropanolic extract of *Cimicifuga racemosa*, 82 postmenopausal women	control group (*n* = 37), isopropanolic extract of *Cimicifuga racemosa* (Remifemin®, 40 mg/day, *n* = 45), duration: 3 months	↓NTX (marker of bone resorption), ↑ ALP (marker of bone formation)	Garcia-Pérez 2009 [59]
Black cohosh (*Cimcifuga racemosa*)	RCT with 3 groups: control (CG), exercise group (EG), exercise and *Cimicifuga racemosa* (CR) supplementation group (EGCR), 128 early postmenopausal women	CG (wellness control, *n* = 42), EG (*n* = 43), EGCR (40 mg/day of CR BNO 1055; *n* = 43), Calcium (1500 mg/d) + vitamin D (500 IE/d) supplementation for all participant duration:12 months	CR (CR BNO 1055) did not enhance positive effects of exercise on BMD at the lumbar spine	Bebenek 2010 [60]

Table 2. *Cont.*

Botanicals	Population and Design	Intervention	Outcome	Authors and References
Labisia pumila and *Eurycoma longifolia*	double-blind RCT with 2 groups: placebo, Nu-femme™, 119 healthy women aged 41–55 years experiencing peri-menopausal or menopausal symptoms	placebo (*n* = 59), herbal formulation (Nu-femme™, *n* = 60) = 200mg Labisia pumila (SLP+®) + 50mg Eurycoma longifolia (Physta®); duration: 24 weeks	No significant effect on bone formation (BALP) and resorption (NTX) markers	Chinnappan 2020 [61]

ALP—alkaline phosphatase, BALP—bone-specific alkaline phosphatase, BMC—bone mineral content, BMD—bone mineral density, CTX—type I collagen crosslinked beta C-telopeptide, DPD—deoxypyridinoline, HRT—hormonal replacement therapy, IGF-1– insulin-like growth factor 1, NTX—type I collagen crosslinked N- telopeptide, OPG—osteoprotegerin, P1NP—type I procollagen-N-propeptide, RANKL—Receptor Activator for Nuclear Factor κB Ligand, SSI—strength strain index, ↑—increased, ↓—decreased

2. Phytoestrogens

Phytoestrogens are naturally occurring nonsteroidal plant compounds that resemble oestrogens and have oestrogenic and/or antiestrogenic activity. They can be divided into two main groups: flavonoids and non-flavonoids. Isoflavones, coumestans, and prenylflavonoids belong to flavonoids, and lignans belong to non-flavonoids [62].

2.1. Isoflavones

Isoflavones are phenolic compounds that belong to the most estrogenic plant-derived substances. Their chemical structure is similar to that of oestradiol. They include, among others, genistein, daidzein, glycitein, biochanin A, and formononetin (Table 3). The main source of isoflavones are legumes belonging to *Fabaceae*: soybean (*Glycine max*) as a source of genistein, daidzein, and glycitein, and red clover (*Trifolium pratense*) as a source of biochanin A and formononetin [62]. In the group of plants containing isoflavones, there are also alfalfa (*Medicago saltiva* L.), beans (green bean, mung bean), psoralea (*Psoralea corylifolia*) and *kudzu* root (*Pueraria lobata* L.) [14]. In the human gastrointestinal tract formononetin, contained in dietary supplements based on red clover, is transformed to daidzein [63]. The amount of isoflavones in soybeans ranges from 1.2 to 4.2 mg per g of dry weight, whereas in red clover, it ranges from 10 to 25 mg per g of dry weight [14]. Isoflavones exert the biologic effect due to two different mechanisms. On the one hand, they act through the classical oestrogen receptor (ER)-mediated signalling pathway, but additionally, it has been described that they may activate intracellular pathways such as protein tyrosine kinase, phospholipase C and MAPK [14]. As most isoflavones are ERβ-selective ligands, it can be supposed that they selectively target bone cells without having an undesired influence on other oestrogen-sensitive tissues, such as the breast and the uterus.

Table 3. Four chemical forms of main isoflavones.

Aglycones	Glycosides	Acetylglycosides	Malonyl Isoflavone Glycosides
Daidzein	Daidzin	Acetyldaidzin	Malonyldaidzin
Genistein	Genistin	Acetylgenistin	Malonylgenistin
Glycitein	Glycitin	Acetylglycitin	Malonylglycitin
Biochanin A	Sissostrin		Malonylsissostrin
Formononetin	Ononin		Malonylononin
Daidzein	Daidzin	Acetyldaidzin	Malonyldaidzin

2.1.1. Soybean in Clinical Trials

The soybean (*Glycine max* L.) is an annual plant belonging to the Fabaceae family, which grows mainly in Southwest Asia. It is a rich source of proteins and flavonoids, such as genistein, daidzein, biochanin A, and glycitein [64]. In soybean, the aglycones and conjugate forms of genistein account for 60% of isoflavones and daidzein for up to 30% [65].

Epidemiological studies have shown that the consumption of food that contains soy may reduce the risk of fracture in postmenopausal women, particularly among those in the early years following menopause [66]. Authors of several observational studies have noticed that populations with a high intake of soy are characterized with a lower incidence of osteoporotic fractures than Western populations [67,68].

To date, many clinical trials (Table 2), systematic reviews, and meta-analyses have been carried out on this topic. Their results suggest that soy phytoestrogens exert significant effects on bone metabolism, and that they inhibit, to some degree, osteoporosis in postmenopausal women [64]. In a study by Scheiber et al., administration of soy isoflavone (60 mg/day) during 12 consecutive weeks increased serum levels of phytoestrogens and ameliorated several key clinical risk factors for osteoporosis in healthy postmenopausal women [27]. Chiechi et al. have showed that supplementation with soya isoflavone in 2 meals twice a week increased bone osteoblastic activity and the serum osteocalcin level [28]. The 12-month-long administration of soy isoflavones (40 mg or 80 mg) maintained hip bone mineral content in later menopause or those with lower body weight or calcium [29]. In another clinical trial, it was reported that isoflavones (126 mg for 6 months) reduced bone loss [30]. The authors suggested that the observed beneficial effect was due to the inhibition of bone resorption in non-obese postmenopausal Chinese women. Similar findings were reported by Wu et al. after supplementation of early postmenopausal women with 75 mg of isoflavone conjugates/day for 1 year [31]. Increased volumetric bone mineral density (vBMD) in postmenopausal women was observed after taking a tablet with isoflavones 80 or 120 mg/day for 3 years [32]. An increase in serum concentrations of bone-specific alkaline phosphatase (BALP) and osteocalcin as markers of increased bone formation were caused by soya isoflavone supplementation (70 mg/day for 12 weeks) in Korean postmenopausal women [33]. Tit et al. reported the similar efficacy of hormonal replacement therapy (HRT) and phytoestrogens in terms of the effects on BMD and bone resorption in postmenopausal women. Two capsules with 40% standardized extract (20 mg soy isoflavones genistein and daidzein per capsule) given orally for 1 year significantly reduced bone resorption [34]. In a randomized clinical trial (RCT) with women during early menopause, Sathyapalan et al. compared the administration of 15 g soy protein with 66 mg isoflavone or 15 g soy protein alone. Moreover, a 6-month long observation revealed that soy reduced bone turnover markers, i.e., type I collagen crosslinked beta C-telopeptide (CTX, bone resorption marker) and type I procollagen-N-propeptide (P1NP, bone formation marker) [35]. However, the results are not consistent, with the study of Levis et al. reporting that supplementation with 200 mg of soy isoflavones daily for 2 years did not protect menopausal women against bone loss [36]. Kreijkamp-Kaspers et al. obtained convergent results—BMD did not differ significantly after the 99 mg supplementation of isoflavones in 25.6 g of soy protein for one year in postmenopausal women [37]. Similarly, another study indicated that the 110 mg/day of soy isoflavone aglycone given for one year in postmenopausal women did not prevent postmenopausal bone loss or affected bone turnover [38].

Phytoestrogen genistein given in the dose 54 mg daily for 1–3 years had positive effects on bone formation and osteopenia in postmenopausal women in several clinical trials [39–41,69]. According to a randomized, placebo-controlled, double-blind study reported by Lappe et al., a lower dose of genistein administered for a shorter time (30 mg daily for 6 months) also prevented osteoporosis development and reduced fracture risk in postmenopausal women [42]. Pawlowski et al. showed that isoflavones mixed in their natural ratios were more effective than genistein-rich soy supplement as bone-preserving agents in postmenopausal women treated for 50 days [70]. Moreover, genistein aglycone in tablets (54 mg daily for 2 years) exerted beneficial effects, not only in postmenopausal osteopenia, but also in women with osteoporosis [43].

The studies mentioned above show that isoflavones ameliorate menopause associated imbalance in bone turnover, protecting BMD and bone strength. These findings suggest that soybean phytoestrogens could be used as a dietary supplement to prevent postmenopausal osteoporosis. Meta-analysis of 63 RCTs found that genistein (54 mg/day) and ipriflavone (600 mg/day) in particular, have beneficial effects on BMD outcomes and are safe in postmenopausal women. Therefore, they may be considered as a complementary or alternative therapy and the prophylaxis of menopause-related osteoporosis [15]. Another meta-analysis of 26 randomized controlled trials (2652 oestrogen-deficient women) found that isoflavones attenuated moderately menopause-associated bone loss in the lumbar spine, femoral neck and distal radius [71]. Additionally, the authors noted that the effect of isoflavones on bone was greater if they were administered as aglycons. The protective influence of soy isoflavones (40–300 mg/day) on osteoporosis-related bone loss and bone mineral density in the femur, neck, lumbar spine and hip was also found in the meta-analysis of 52 controlled trials (5313 patients) [72]. However, the effectiveness of soy isoflavone supplementation in treatment and prophylaxis of osteoporosis in peri- and postmenopausal females remain debatable. In a systematic review of nine studies (1379 women), Perna et al. found no consensus regarding the protective effect of soy isoflavones (20–80 mg/day) on bone loss. However, the authors did not exclude the possible protective effect of soy isoflavones on bone metabolism. Similar conclusions of a systematic review of 23 clinical trials were reported by other authors that found only a minimal effect of isoflavones on bone mineral density in menopausal women [73]. Several other meta-analyses reported that the effects were minimal [43,74,75] or none, as mentioned above [36]. The antiosteoporotic effects of flavonoids seem to depend on the balance between their estrogenic agonist and antagonist properties [76]. Their beneficial influence on bone metabolism may also be derived from their other biochemical properties, including enzymatic inhibition of certain protein kinases or activation of estrogen type I receptors [64]. Some authors indicate that equol—an isoflavandiol produced by gut microflora from daidzein and possessing a higher estrogenic activity than the predominant flavonoids—may be responsible for the clinical effectiveness of flavonoids [77]. The discrepancies between the results of the reported studies may also be attributed to differences in the study design.

2.1.2. Red Clover in Clinical Trials

Red clover (*Trifolium pratense* L.) belongs to the legume family and is often used to relieve symptoms of menopause, high cholesterol, and osteoporosis [78]. Isoflavones: biochanin A, formononetin, and sissotrin, are responsible for its estrogenic activity. In intestines, biochanin A and formononetin are demethylated and metabolized to genistein and daidzein [79]. The bone-preserving effects of red clover have also been examined, but not as extensively as those of soy [80].

In a randomized, placebo-controlled study, an isoflavone preparation (Rimostil®) containing genistein, daidzein, formononetin and biochanin A was administered to 46 postmenopausal women in a double-blind protocol after a single-blind placebo phase, and followed by a single-blind washout phase. A 6-month-long administration of an isoflavone combination extracted from red clover (57 mg/day or 85.5 mg/day) to postmenopausal females increased the BMD of radius and ulna [44]. In another clinical trial ($n = 205$), the red clover extract containing 41 mg isoflavone per tablet (Promensil®) ameliorated the decrease of bone mineral content (BMC) and BMD in lumbar spine in pre-, peri-, and postmenopausal women taking the supplement for 12 months. Authors also reported the elevation of bone formation markers [45].

In another 12-month, double-blind, parallel design RCT, 78 postmenopausal osteopenic women were supplemented with calcium (1200 mg/day), magnesium (550 mg/day), calcitriol (25 mg/day) and given either red clover extracts rich in isoflavone aglycones and probiotics (RCE, 60 mg isoflavone aglycones/day and probiotics) or a masked placebo. RCE intake combined with supplementation (calcium, magnesium, and calcitriol) was more effective than supplementation alone. Twice daily RCE intake over one year prevented a menopause-associated decrease of BMD normalizing bone turnover, promoting a favourable oestrogen metabolite profile (2-OH:16α-OH), and stimulating equal production in postmenopausal women with osteopenia [46].

Thorup et al. found that the intake of 150 mL red clover extract containing 37.1mg isoflavones for 12 weeks improved bone health in menopausal women (n = 60). The conclusions were based on BMD and T-score at the lumbar spine and plasma CTX levels [47]. However, a review of the potential skeletal benefit of red clover concluded that there was limited evidence of efficacy [81]. For example, in a placebo-controlled 3-year trial in 401 women with a family history of breast cancer, 40 mg of red clover produced no effect on BMD [48].

In another study with perimenopausal women (n = 250), when taking two tablets per day containing red clover extract (28.6 mg or 41 mg isoflavones) for 12 weeks, no significant differences in bone turnover markers were observed compared to placebo [49].

Although the evidence is limited, it appears that red clover isoflavones may have a beneficial effect on bone mineral density in peri- and postmenopausal women [80].

2.1.3. Soybean and Red Clover in Animal Studies and In Vitro Models

The studies that investigated the effects of soybean on markers of bone turnover in ovariectomized rats reported contradictory results. Park et al. reported that soybean increased serum osteocalcin levels and decreased urinary deoxypyridinoline (DPD) levels [82], while Byun et al. observed a decrease in osteocalcin and DPD levels [83]. However, other authors detected no influence of soybean on bone turnover markers in ovariectomized rats [84]. Hinton et al. reported that soybean improved whole bone and tissue level biomechanical properties in ovariectomized rats, although it did not improve the trabecular microarchitecture [84].

Soybean proteins contain a high level of phytate, which decreases calcium bioavailability [85], therefore, the investigations assessing the phytate-removed soybean proteins of bone metabolism were conducted. Phytate-removed and deamidated soybean β-conglycinin enhanced calcium absorption from the intestines in ovariectomized rats [85]. As a consequence, an increase in serum calcium level normalized PTH secretion. Suppression of ovariectomy-induced changes in bone turnover was also observed. Additionally, Akao et al. reported a reduction of bone resorption, enhanced BMD, and strengthened bone in ovariectomized rats receiving phytate-removed and deamidated soybean β-conglycinin [85]. However, the influence on trabecular BMD was less prominent than the influence of cortical BMD.

Soy isoflavones bind to ERβ that are expressed in the calcaneus but not in cortical bone [7]. This fact explains why they mainly influence the trabecular bone. In vivo, soy isoflavones through Smad's activation in osteoblasts lead to the upregulation of the expression of Runx2 and OSX that are important transcription factors involved in osteoblast differentiation and proliferation [86]. Soy isoflavones decreased RANKL levels. They increase the expression of OPG, β-catenin, and Wnt 3a and 7b in osteoblasts. Noh et al. reported that the combination of soy isoflavone and hop prenylflavones (Soy-Hop) had a protective influence on bone in ovariectomized rats [87]. In their study, Soy–Hop administration in a dose-dependent manner reduced ovariectomy-induced elevation of osteocalcin, alkaline phosphatase (ALP), and CTX levels. It also attenuated the ovariectomy-induced expression of RANKL messenger ribonucleic acid (mRNA). A micro-computed tomography (mCT) examination revealed reduced porosity and decreased separation between trabeculae in the femoral epiphysis in Sop-Hop receiving ovariectomized rats [87]. Kim et al. reported that dry-fermented soybean food ameliorated senile osteoporosis in the senescence-accelerated mouse prone 6 (SAMP6) model [88].

In vitro studies demonstrated that daidzein and genistein bound to RANKL within the side residues involved in RANK binding [89] prevented the formation of the complex of RANKL-RANK that activates bone resorption. Additionally, it was demonstrated that soy isoflavones increased Runx2 expression mineralisation in human osteosarcoma Saos-2 cell culture that activated osteoblasts and led to the acceleration of matrix mineralization [89].

Genistein was also shown to be able to elevate ALP activity and decrease RANKL/OPG ratio in Saos-2 [90]. There are data matching the activation of osteoblasts by genistein with its binding to oestrogen receptor β present on osteoblastic cells [91,92]. Genistein is twenty times more selective for oestrogen receptor β (ERβ) than α [93]. Animal studies confirmed that genistein combined with silicon and zinc significantly reduced RANKL expression and prevented ovariectomy-induced bone resorption [94,95]

Daidzein is the most widely studied soy phytoestrogen. Daidzein was also reported to stimulate osteoblast differentiation. It stimulates osteoblasts through the BMP-2/Smad/Runx2 pathway [96]. It was reported that oestrogen receptor signalling, mitogen-activated protein kinase/extracellular signal-regulated kinases (MAPK/ERK), and phosphoinositide-3-kinase/serine-threonine protein kinase B (PI3K/AKT) were involved in osteoblast activation via daidzein [97].

The summary of the influence of isoflavones on bone metabolism is presented in Figure 3.

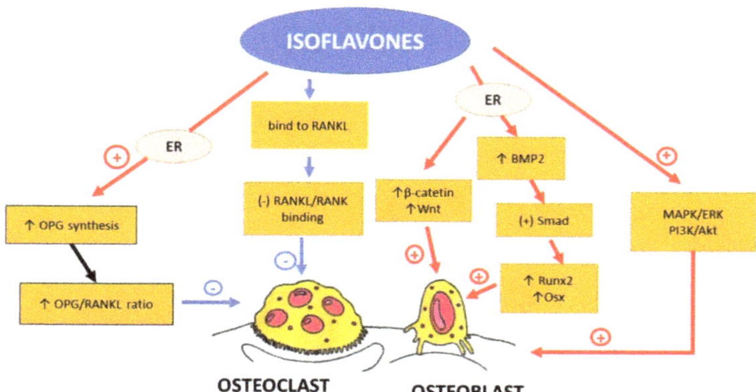

Figure 3. Schematic diagram representing the influence of isoflavones on proliferation, differentiation and activity of osteoclasts and osteoblasts. BMP2—bone morphogenic protein 2, ER—oestrogen receptor, MAPK/ERK—mitogen-activated protein kinase/extracellular signal-regulated kinases signalling pathway, OPG—osteoprotegerin, Osx—Osteoblast-specific transcription factor Osterix, PI3K/Akt-phosphoinositide-3-kinase/serine-threonine kinase signalling pathway, —Receptor Activator for Nuclear Factor κB, RANKL—Receptor Activator for Nuclear Factor κB Ligand, Runx2—Runt-related transcription factor 2, ↑—increased, ↓—decreased, (-)—inhibited, (+)—activated

2.1.4. Other Plants Containing Isoflavones

Alfalfa

Alfalfa (*Medicago sativa* L.), also called lucerne, belongs, as red clover, to the legume family. It is cultivated as a forage crop in many countries over the world. Its sprouts are a common ingredient of dishes made in Indian cuisine. Alfalfa, as other legumes, is a known source of phytoestrogens, including spinasterol, coumestrol, coumestan, and ipriflavone. As mentioned above, the meta-analysis of 63 controlled trials investigating 6427 postmenopausal women revealed that ipriflavone (600 mg/day) is a promising molecule for the prevention and treatment of postmenopausal osteoporosis [15]. Ipriflavone has been reported to induce osteoblast proliferation and prevent menopause-related bone loss.

Pueraria candollei var. mirifica

Pueraria candollei var. mirifica (Airy Shaw and Suvat.) Niyomdham (commonly termed *P. mirifica*), known also as *kudzu* root, has a long history as a postmenopausal rejuvenate therapy for indigenous people. It contains various isoflavones: puerarin, daidzein, daidzin, mirficin and salvianololic acid. In a double-blind RCT of healthy postmenopausal women aged 45 to 60 years old, Manonai et al. showed that *Pueraria mirifica* at a dose of 20, 30, and 50 mg/day for a 24-week period demonstrated an oestrogen-like effect on bone turnover. The P1NP level was reduced as seen with other treatments with Erα agonists [50]. In another double-blind RCT, nineteen postmenopausal women (12/7 test/control) received *P. mirifica* powder or placebo for 2 months. Investigators also found a reduced ALP and commented on its relationship to bone preservation, but the isoform targeted was not stated [51].

Pueraria extract prevented ovariectomy-induced bone loss in rats [98]. Puerarin, that is, the main active ingredient of *Pueraria* extract, slows down the bone loss and reverses the ovariectomy-induced increase in bone turnover in rats [99]. It also alleviated osteopenia and prevented the deterioration of trabecular structure in mCT [99]. Other authors reported that puerarin inhibited RANKL-dependent osteoclastogenesis [100], induced mineral nodulus formation in osteoblasts through the activation of PI3K/AKT signalling pathway [101] and promotes osteoblast differentiation [102].

Summing up, isoflavones not only prevent bone resorption by the inhibition of RANKL/RANK interaction and osteoclast maturation and differentiation, but they also seem to promote bone formation. They increase, among others, the expression of BMP2 and Runx2, that are involved in the activation and differentiation of osteoblasts. The molecular mechanism of phytoestrogen influence on bone metabolism is very complex and there are many possible pathways that might be involved (Figure 3).

2.2. Other Plants Containing Phytoestrogens Investigated in Osteoporosis Treatment

2.2.1. Epimedium (Berberidaceae)

Epimedium in Clinical Trials

Epimedium is a genus of about 52 species in the family *Berberidaceae*, which is also known as Rowdy Lamb Herb, Xianlinpi, Barrenwort, Bishop's Hat, Fairy Wings, Horny Goat Weed, and Yangheye or Yin Yang Huo). The traditional Chinese medicinal herb Epimedii has been utilized for centuries to treat bone fractures, bone loss, and menopause-associated disorders [64]. The results of recent clinical trials have reported suggest that compounds or extracts of Epimedium may prevent or delay the onset of osteoporosis and reduce the risk of hip fractures [21]. Icariin is a prenylated flavonol glycoside isolated from Epimedium herbs, and has been shown to be the main bioactive component [16]. In clinics, Epimedium is used to treat osteoporosis, climacteric period syndrome, breast lumps, hyperpiesia, and coronary heart disease [103].

In a 24-month double-blind RCT in healthy, late postmenopausal women, the intervention group ($n = 50$, a daily dose of 60 mg icariin, 15 mg daidzein, and 3 mg genistein) had a significantly reduced bone loss compared to the placebo group ($n = 50$). Treatment with icariin maintained BMD at 12 months. A long-term (up to 12–24 months) administration of icariin improved BMD in the lumbar spine and femoral neck in a time-dependent manner. Although the effect of icariin is less effective in the improvement in BMD than oestrogen replacement or treatment with bisphosphonates, it seems to be an attractive alternative therapy due to its low risk of severe side effects. It exerted no oestrogenic effect on the uterus and did not change the serum estradiol level, proving its safety when it comes to the endometrium. A 2-year-long treatment with icariin was also not associated with the incidence of breast cancer or cardiovascular events [52]. Further clinical trials encompassing a larger population are needed to investigate the influence of icariin and its derivatives on bone formation and regeneration in humans, as well as its safety profile [16].

Epimedium in Animal Models and In Vitro Studies

Epimedium flavonoids (icariin, epimedin B, and epimedin C), that possess oestrogenic activity, have been identified as the main constituents of Epimedium plants that exert antiosteoporotic activity, as they inhibit bone resorption, promote bone formation and block urinary calcium excretion [21]. The flavonoids from *Epimedium* promote osteoblast activity through the regulation of the expression of IL-6 (interleukin 6), OPG, RANKL, M-CSF, BMP-2, and Smad4. They modulate the BMP/Smad4 and Wnt/β-catenin signalling pathways, inducing osteoblast differentiation [104]. Icariin is the most abundant flavonoid in *Herba Epimedii* and has a better antiresorptive effect than other components isolated from *Epimedium* plants. It stimulates bone formation by the promotion of osteoblasts differentiation and the enhancement of their activity [16,105]. Icariin activates BMP-2/Smad4, Wnt and IGF-1 signal transduction pathways [5,17], induces ERK (extracellular signal-regulated kinase), JNK (c-Jun N terminal kinase) and p38 kinase activation [18]. Icariin not only promotes bone formation, but also inhibits osteoclast differentiation and bone resorption. It decreases RANKL-induced osteoclastogenesis via the modulation of NFκB and MAPK expression and downregulation of main regulators of osteoclastogenesis (c-fos and NFAT c1—nuclear factor of activated T-cells, cytoplasmic 1) [19]. Micro-CT results suggest that icariin improves the bone parameters (BMD, bone volume/tissue volume—BV/TV, connectivity density—Conn.D) and restores bone structure in ovariectomized animals [106]. Ikarisoside A, a flavonoid isolated from *Epimedium koreanum*, also inhibits RANKL-induced osteoclastogenesis [104].

2.2.2. Hop (*Humulus lupulus* L.)

Hop (*Humulus lupulus* L.), which belongs to the Cannabaceae family, has been used worldwide in the brewing industry as a source of bitterness in beer. Apart from this, hop extract is known for containing phytoestrogen components and exerting oestrogenic effects. In general, compounds of the oestrogenically active fraction of lupulin gland secretion belong in the following prenylflavonoids: xanthohumol, being the most abundant prenylflavonoid in hops, izoxanthohumol, 6-prenylnaringenin and 8-prenylnaringenin [107]. Moreover, 8-prenylnaringenin has stronger oestrogenic properties than soy phytoestrogens [108]. Ban et al. reported that hop extract Lifenol® prevented osteoporosis development in ovariectomized rats [109]. Hop extract ameliorated the ovariectomy-induced decreased of BMD, femur weight, and BMC (bone mineral content). Additionally, it restored the trabecular structure of calcaneus bone and inhibited ovariectomy-induced osteoclast activation. A mild osteoprotective effect of hop extract was also reported by other authors [110]. Li et al. reported that xanthohumol blocks RANKL-induced osteoblast differentiation and bone resorption, in vitro and in vivo, in ovariectomized mice [111]. At the molecular level, it blocks the RANKL/TRAF6 (tumour necrosis factor receptor associated factor 6) signalling pathway involved in osteoclastogenesis. Additionally, xanthohumol stimulates osteogenic marker gene expression in mesenchymal and pre-osteoblastic cells [112]. Furthermore, 8-prenylnaringenin, that is, the strongest phytoestrogen known, similarly to soy phytoestrogen, exerts its osteoprotective effect through ERs. It inhibits RANKL expression and induces the expression of ostoprotegerin (OPG), which is an inhibitor of osteoclast activity [113].

3. Other Botanicals

3.1. Dried Plums

3.1.1. Dried Plums in Clinical Trials

In a rat model of ovariectomy-induced osteoporosis, dried plum (*Prunus domestica* L.) prevented the bone loss and structural damage of bone tissue [114]. The studies described below have tried to confirm this effect in humans, mainly in osteopenic postmenopausal women.

Three-month RCT comparing the influence of dried plums versus dried apples on biomarkers of bone formation in 58 postmenopausal women has revealed that the consumption of 100 g/day dried plums significantly increased the serum markers of bone formation: total ALP, bone-specific ALP (BALP) and IGF-1 [53]. Another one-year RCT compared the effects of daily consumption of 100 g dried plum to 75 g dried apple (control) on BMD and biomarkers of bone turnover in 160 osteopenic postmenopausal women. Hooshamnd et al. reported that dried plum improved lumbar and ulnar BMD when compared to dried apples [115]. Additionally, the authors reported that dried plum increased RANKL and OPG concentration, and decreased serum sclerostin level, however the reported results did not reach statistical significance [54].

Similarly, inconsistent results were obtained in non-randomized six-month intervention trials evaluating the effects of resistance training with and without dried plum at a dose of 90 g in 23 postmenopausal breast cancer survivors. In both groups, an improvement of upper and lower body strength was found, but no improvements in body composition or BMD was detected [116]. However, in a subsequent six-month clinical trial evaluating the efficacy of two doses of dried plum (50 g vs. 100 g) in 48 older postmenopausal women, it was reported that dried plums prevented the loss of total body BMD and reduced the serum concentration of tartrate-resistant acid phosphatase 5b (TRAP-5b). Additionally, the authors concluded that both doses of dried plaums are equally effective [55]. The beneficial effect was also observed in the trial, with 35 men between the ages of 55 and 80 with moderate bone loss. Patients were randomized into one of three groups: 100 g dried plum daily, 50 g dried plum daily, or control group. All three groups also consumed a multivitamin containing 450 mg calcium and 800 IU vitamin D. After three months, decreased serum concentration of osteocalcin was observed, as well as an elevation of OPG/RANKL. Authors suggested that regular consumption of either 100 g or 50 g dried plum for three months may make some contributions to bone formation and bone turnover activity, and a minimal contribution to decreasing inflammation and improving bone density and quality [56].

The results of the presented studies suggest that dried plum is a promising functional food therapy for preventing bone loss, with the potential for long-lasting bone protective effects [114].

3.1.2. Dried Plums in Animal Studies and In Vitro Models

Dried plums contain carbohydrates, vitamins A, B and K, potassium, calcium, magnesium, boron, selenium, dietary fibres, and polyphenols, including chlorogenic acid, rutin and proanthocyanidin [117]. Animal studies comparing the influence of dried plums and standard diet on bone metabolism and bone mechanical properties showed that diet supplementation with dried plums increased vertebral and femoral bone mineral density [118,119]. In ovariectomized animals, the administration of dried plums increased bone mineral density and the number of trabeculae (Tb.N.), and decreased the separation of trabeculae (Tb.Sp.) [120,121]. Further animal studies revealed that polyphenols are the main bioactive compounds responsible for bone response to therapy with dried plums. However, the addition of potassium and vitamin K to the polyphenolic resulted in the additional increase of bone mineral density [122]. In in vitro studies, dried plum polyphenols suppressed osteoclast activity and differentiation [123], increased mineral nodule formation and osteoblast activity [124].

3.2. Horsetail (Equisetum arvense)

Horsetail (*Equisetum arvense* L.) is widely distributed over the northern hemisphere. Extracts and other preparations of horsteil have been used for ages in European folk medicine. It contains abundant constituents that may exert beneficial effects on bone health, e.g., silica, flavonoids, and triterpenoids.

The only clinical study evaluating the effectiveness of horsetail in the treatment of perimenopausal osteoporosis recruited 122 women in menopause for at least two years, who had not undergone oestrogen replacement therapy or drug therapy for recalcification: 30 patients were administered with titrated dry horsetail extract for 80 days; 31 patients were administered with placebo for 40 days and titrated horsetail extract for a further 40 days; 29 patients received no treatment whatsoever; 32 patients were treated with Osteosil Calcium for 80 days. All patients received two tablets per day according to procedures for randomized double-blind studies. Patients who received treatment with titrated horsetail extract after the period of placebo administration showed the same changes observed in patients treated with the active ingredient; treatment with titrated horsetail extract and with Osteosil Calcium improved bone metabolism and BMD [57].

E. arvense has a high concentration of silica, and it has been demonstrated in vitro that the horsetail extract induced bone regeneration [125] and inhibited osteoclastogenesis [126]. It has been reported that horsetail extract enhanced bone mineralization and bone formation in ovariectomized rats [127]. Additionally, a diet containing horsetail extract (120 mg/kg) increased bone mineral density in rats [128]. However, there are scarce studies to support the hypothesis of the beneficial effects of horsetail on bone health, and the European Food Safety Authority concluded that there is not enough evidence of the bone-protecting influence of *E. arvense* [129].

3.3. Black Cohosh (Cimcifuga racemosa)

Data from the following clinical trials suggest the beneficial effects of *Cimicifuga racemosa* on bone metabolism and bone mineral density. Additionally, the authors hint at the possible reduction of the cumulative dose of HRT for the prophylaxis of osteoporosis in patients receiving CR [130].

A double-blind RCT on postmenopausal women showed that CR stimulated osteoblast activity, and improved markers of bone turnover [58]. Other authors reported that *C. racemosa* extract reduced bone resorption (decrease in the urinary level of N-telopeptide) and increased bone formation (elevation of ALP) in postmenopausal women. However, serum obtained from treated females did not stimulated osteoblasts' culture, but failed to demonstrate a direct stimulating effect of the serum from treated women on a culture of osteoblasts [59]. On the other hand, other authors did not find a bone-favourable effect of *C. racemosa* extract in exercising early postmenopausal women [60]. The absence of a *C. racemosa* -taking non-exercising comparison group was a significant shortcoming of this study, as the possibly positive effect of *C. racemosa* might have been lost in the well-known considerable favourable effect of exercise on BMD [131]. Another trial measuring serum osteocalcin and C-terminal telopeptide [132] did not reveal any significant difference between black cohosh and placebo in measured outcome at 12 weeks.

In animal studies, *Cimicifuga racemosa* increased BMD and restored bone architecture (preventing the decline in BV/TV, Tb.Th., and Tb.N., and preserving SMI—Structural Model Index) in ovariectomized animals [133]. Cycloartane, a tripentoid glycoside isolated from black cohosh, inhibits NFκB and ERK signalling pathways that leads to inhibition of RANKL-induced osteoclast differentiation [134]. Additionally, actein and deoxyactein protect osteoblasts against oxidative stress and promote cell growth and matrix mineralisation [135,136].

3.4. Salvia miltiorrhiza and Salvia plebia

Red sage (*Salvia miltiorrhiza Bunge*), also known as Danshen in Chinese, has been used to treat bone diseases in traditional Chinese medicine. Guo et al. analyzed clinical trials that investigated the efficacy of *Salvia miltiorrhiza* in the treatment of osteoporosis. In reported trials *S. miltiorrhiza* was given as monotherapy or as a part of combined therapy with other plants or ingredients. They identified 36 trials that demonstrated high efficacy and no toxicity of *S. miltiorrhiza*, however, in some studies, small patient samples, short treatment duration, frequent lack of detailed numerical data, and no clear endpoints limited their value and reliability [137]. *S. miltiorrhiza* influence on bone regeneration was also investigated in patients with avascular and ischemic necrosis of femoral head. *S. miltiorrhiza* was injected and implantated in the calcium phosphate cement/*S. miltiorrhiza* drug delivery system by minimal invasive surgery. The digital substruction arterography and X-ray films demonstrated that *S. miltiorrhiza* administration improved the microcirculation and regeneration of the affected bone [138].

In animal studies, *S. miltiorrhiza* and *S. plebia* prevented ovariectomy-induced decrease in trabecular bone mass and BMD. It also reduced TRAP5b activity and oxidative stress in ovariectomised animals [22,26,139]. Tashinones, salvianolic acid, and eudebeiolide B have been identified as osteoprotective components of Salivia plants. Tanshinones inhibit the formation of TRAP5b-expressing osteoclasts by suppressing the RANKL-induced expression of c-fos and NFATc1 [22,23]. Salvianolic acid A and B modulate osteoblast differentiation and upregulate osteoblast activity [24,25]. Liu et al. reported that *Radix salviae* improves bone microarchitecture and biomechanical properties through the Wnt/β-catening signalling pathway in ovariectomized rats [140].

3.5. Other Herbs

3.5.1. *Labisia pumila* and *Eurycoma longifolia*

Labisia pumila that belongs to the family *Myrsinaceae* is used in Asia for the treatment of painful menstruation and disorders of sexual life in females due to its oestrogenic properties. As a phytoestrogen-containing plant, it is also used in osteoporosis treatment [141].

Both *L. pumila* and *E. longifolia* have demonstrated a protective effect on bone loss due to osteoporosis in previously published studies. In a double-blind, 24-week RCT, 119 healthy women (aged 41–55 years) experiencing peri-menopausal or menopausal symptoms were enrolled and supplemented with herbal formulation (Nu-femme™) comprising LP (SLP+®) and *Eurycoma longifolia* (Physta®) or placebo. There were no significant differences between- and within-group of bone markers for osteoporosis reflecting bone formation (BALP) and resorption (N-terminal telopeptide—NTX) [61].

3.5.2. *Drynaria fortunei*

Rhizoma Drynariae, the dried rhizome of *Drynaria fortunei (Kunze) J. Sm.*, is reported to prevent age-associated bone loss. It contains mainly flavonoids, triterpenoids, phenolic acids, and glycosides [142]. In ovariectomized animals, *Rhizoma Drynariae* extract prevented oestrogen deficiency-induced weight gain without an unfavourable effect on the uterus [143]. Additionally, it exerted a protective effect on bone, increasing Tb.N. and bone fraction (BV/TV), and decreased Tb.Sp. in calcaneus bone. In vitro studies revealed that *Rhizoma Drynariae* extract inhibits RANK activity [143]. Sun et al. reported that polysaccharides from *Rhizoma Drynariae* exerts an antiosteoporotic effect in ovariectomized rats. It maintained trabecular microarchitecture and bone biomechanical properties, and increased femoral and tibial bone mineral density (BMD) [144].

3.5.3. Other Plant-Derived Constituents

As there is a great need to develop new drugs that might be used in the treatment of osteoporosis, there are plenty of reports on studies investigating the influence of plant-derived bioactive substances on the activity and differentiation of osteoclasts and osteoblasts. Loureirin B (flavonoid, extracted from *Dracaena cochinchinensis*) and kirenol (diterpenoid extracted from the Chinese herbal medicine *Siegesbeckiae*) inhibit RANKL-induced osteoclast differentiation by attenuation of NFAT expression [145,146]. Kaempferol (natural flavonol found in various plants, e.g., tea and broccoli) that exerts oestrogenic properties, on the one hand, inhibits bone resorption and on the other promotes bone formation [147]. Its bone-protective effect is mediated through regulation of oestrogen receptor, bone morphogenetic protein-2 (BMP-2), NF-κB, MAPK and mammalian target of rapamycin (mTOR) signalling pathways [148]. Mangiferin (xanthone originally extracted from mango tree) attenuates ovariectomy-induced osteoporosis in rats [149] and promotes osteoblast differentiation through the increased expression of Runx2 and BMP2/Smad1 signalling pathway [150,151]. Quercetin (flavanol wildly distributed in plants, e.g., red onion) inhibits RANKL-mediated osteoblastogenesis through Wnt, NFκB, Nrf2 (nuclear factor erythroid 2-related factor 2), and SMAD-dependent signalling pathways [152].

4. Conclusions

Traditional folk medicine is a rich source of bioactive compounds waiting for discovery and investigation that might be used in treatment and prophylaxis od osteoporosis. The mechanism of action of some chosen botanicals are presented in Figure 4 and Table 2.

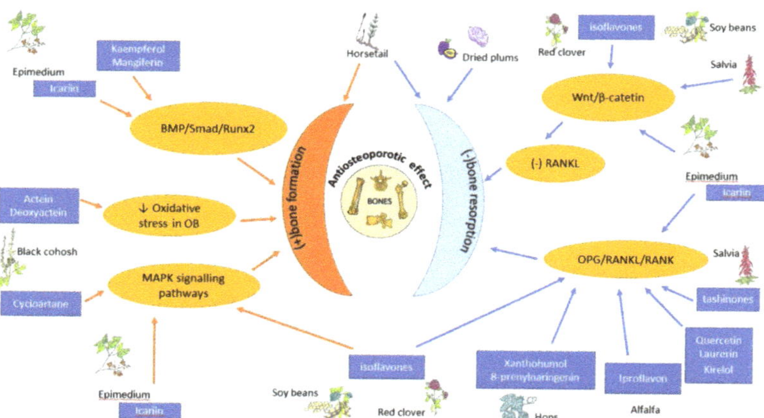

Figure 4. Schematic diagram of the antiosteoporotic activity of various plants and their components. BMP—bone morphogenic protein, MAPK—mitogen-activated protein kinase, OB—osteoblast, OPG—osteoprotegerin, RANK—Receptor Activator for Nuclear Factor κB, RANKL—Receptor Activator for Nuclear Factor κB Ligand, Runx2—Runt-related transcription factor 2, ↓ decreased, (-)—inhibited, (+) activated

Author Contributions: Conceptualization, W.S. and B.N.; methodology, B.N.; resources, W.S. and B.N.; original draft preparation, W.S., P.J. and B.N.; writing—review and editing, B.N.; visualization, P.J. All authors have read and agreed to the published version of the manuscript.

Funding: This research received no external funding.

Institutional Review Board Statement: Not applicable.

Informed Consent Statement: Not applicable.

Data Availability Statement: No new data were created or analyzed in this study. Data sharing is not applicable to this article.

Conflicts of Interest: The authors declare no conflict of interest.

Abbreviations

AKT	protein kinase B
ALP	alkaline phosphatase
BALP	bone-specific alkaline phosphatase
bFGF	basic fibroblast growth factor
BMC	bone mineral content (BMC)
BMD	bone mineral density
BMP	bone morphogenic protein
BMSC	bone marrow-derived mesenchymal stem cells
BV/TV	bone fraction
c-FMS	colony-stimulating factor-1 receptor
Conn.D.	connectivity density
CTX	type I collagen crosslinked beta C-telopeptide
DPD	deoxypyridinoline
ER	oestrogen receptor
ERK	extracellular signal-regulated kinases
FasL	Fas ligand

bFGF	basic fibroblast growth factor
Hh	Hedgehog
HRT	hormonal replacement therapy
IGF	insulin-like growth factor
IL-6	interleukin 6
JNK	c-Jun N terminal kinase
MAPK	mitogen-activated protein kinase
M-CSF	macrophage colony-stimulating factor
mCT	micro-computed tomography
mRNA	messenger ribonucleic acid
mTOR	mechanistic target of rapamycin
NFAT-c1	Nuclear factor of activated T-cells, cytoplasmic 1
NFκB	nuclear factor-kappa B
Nrf2	nuclear factor erythroid 2-related factor 2
NTX	type I collagen crosslinked N- telopeptide
OB	osteoblast
OC	osteoclast
OPG	osteoprotegerin
OSX	Osterix
P1NP	type I procollagen-N-propeptide
PDGF	platelet-derived growth factor
PI3K/AKT	phosphoinositide-3-kinase/serine-threonine protein kinase B
PLC	phospholipase C
pre-OB	pre-osteoblasts
pre-OC	pre-osteoclasts
PTH	parathyroid hormone
RANK	Receptor Activator for Nuclear Factor κB
RANKL	Receptor Activator for Nuclear Factor κB Ligand
RCT	randomized clinical trial
Runx2	Runt-related transcription factor 2
SERM	Selective Oestrogen Receptor Modulator
sFasL	soluble Fas lignad
SMI	Structural Model Inde
Tb.N.	number of trabeculae
Tb.Sp.	separation of trabeculae
Tb.Th.	trabecular thickness
TGFβ	tumour growth factor β
TRAF6	tumour necrosis factor receptor associated factor 6
TRAP 5b	Tartrate-resistant acid phosphatase 5b

References

1. Consensus development conference: Diagnosis, prophylaxis, and treatment of osteoporosis. *Am. J. Med.* **1993**, *94*, 646–650. [CrossRef]
2. Bartl, R.; Bartl, C. Epidemiology of osteoporotic fractures. In *The Osteoporosis Manual*; Springer International Publishing: Cham, Switzerland, 2019; pp. 231–232.
3. Hernlund, E.; Svedbom, A.; Ivergård, M.; Compston, J.E.; Cooper, C.; Stenmark, J.; McCloskey, E.V.; Jonsson, B.; Kanis, J.A. Osteoporosis in the European Union: Medical management, epidemiology and economic burden. *Arch. Osteoporos.* **2013**, *8*, 136. [CrossRef] [PubMed]
4. Kanis, J.; Cooper, C.; Rizzoli, R.; Reginster, J.Y.; On behalf of the Scientific Advisory Board of the European Society for Clinical and Economic Aspects of Osteoporosis (ESCEO) and the Committees of Scientific Advisors and National Societies of the International Osteoporosis Foundation (IOF); Cooper, C.; Rizzoli, R.; Reginster, J.-Y. European guidance for the diagnosis and management of osteoporosis in Postmenopausal women. *Osteoporos. Int.* **2019**, *30*, 3–44. [CrossRef]
5. Bellavia, D.; Dimarco, E.; Costa, V.; Carina, V.; De Luca, A.; Raimondi, L.; Fini, M.; Gentile, C.; Caradonna, F.; Giavaresi, G. Flavonoids in Bone Erosive Diseases: Perspectives in Osteoporosis Treatment. *Trends Endocrinol. Metab.* **2021**, *32*, 76–94. [CrossRef]
6. Cauley, J.A. Estrogen and bone health in men and women. *Steroids* **2015**, *99*, 11–15. [CrossRef]
7. Khalid, A.B.; Krum, S.A. Estrogen receptors alpha and beta in bone. *Bone* **2016**, *87*, 130–135. [CrossRef]
8. Krum, S.A.; Chang, J.; Miranda-Carboni, G.; Wang, C.-Y. Novel functions for NFκB: Inhibition of bone formation. *Nat. Rev. Rheumatol.* **2010**, *6*, 607–611. [CrossRef] [PubMed]

9. Garcia, A.J.; Tom, C.; Guemes, M.; Polanco, G.; Mayorga, M.E.; Wend, K.; Miranda-Carboni, G.A.; Krum, S.A. ERα signaling regulates MMP3 expression to induce FasL cleavage and osteoclast apoptosis. *J. Bone Miner. Res.* **2013**, *28*, 283–290. [CrossRef]

10. Martin, A.; Xiong, J.; Koromila, T.; Ji, J.S.; Chang, S.; Song, Y.S.; Miller, J.L.; Han, C.-Y.; Kostenuik, P.; Krum, S.A.; et al. Estrogens antagonize RUNX2-mediated osteoblast-driven osteoclastogenesis through regulating RANKL membrane association. *Bone* **2015**, *75*, 96–104. [CrossRef]

11. Gazzerro, E.; Canalis, E. Bone morphogenetic proteins and their antagonists. *Rev. Endocr. Metab. Disord.* **2006**, *7*, 51–65. [CrossRef]

12. Canalis, E. Wnt signalling in osteoporosis: Mechanisms and novel therapeutic approaches. *Nat. Rev. Endocrinol.* **2013**, *9*, 575–583. [CrossRef] [PubMed]

13. Canalis, E. Skeletal Growth Factors. In *Osteoporosis*; Elsevier Academic Press: Cambridge, MA, USA, 2013; pp. 391–410.

14. Gómez-Zorita, S.; González-Arceo, M.; Fernández-Quintela, A.; Eseberri, I.; Trepiana, J.; Portillo, M.P. Scientific Evidence Supporting the Beneficial Effects of Isoflavones on Human Health. *Nutrients* **2020**, *12*, 3853. [CrossRef]

15. Sansai, K.; Na Takuathung, M.; Khatsri, R.; Teekachunhatean, S.; Hanprasertpong, N.; Koonrungsesomboon, N. Effects of isoflavone interventions on bone mineral density in postmenopausal women: A systematic review and meta-analysis of randomized controlled trials. *Osteoporos. Int.* **2020**, *31*, 1853–1864. [CrossRef]

16. Wang, Z.; Wang, D.; Yang, D.; Zhen, W.; Zhang, J.; Peng, S. The effect of icariin on bone metabolism and its potential clinical application. *Osteoporos. Int.* **2018**, *29*, 535–544. [CrossRef] [PubMed]

17. Liang, W.; Lin, M.; Li, X.; Li, C.; Gao, B.; Gan, H.; Yang, Z.; Lin, X.; Liao, L.; Yang, M. Icariin promotes bone formation via the BMP-2/Smad4 signal transduction pathway in the hFOB 1.19 human osteoblastic cell line. *Int. J. Mol. Med.* **2012**, *30*, 889–895. [CrossRef] [PubMed]

18. Song, L.; Zhao, J.; Zhang, X.; Li, H.; Zhou, Y. Icariin induces osteoblast proliferation, differentiation and mineralization through estrogen receptor-mediated ERK and JNK signal activation. *Eur. J. Pharmacol.* **2013**, *714*, 15–22. [CrossRef]

19. Xu, Q.; Chen, G.; Liu, X.; Dai, M.; Zhang, B. Icariin inhibits RANKL-induced osteoclastogenesis via modulation of the NF-κB and MAPK signaling pathways. *Biochem. Biophys. Res. Commun.* **2019**, *508*, 902–906. [CrossRef]

20. Ming, L.-G.; Lv, X.; Ma, X.-N.; Ge, B.-F.; Zhen, P.; Song, P.; Zhou, J.; Ma, H.-P.; Xian, C.J.; Chen, K.-M. The Prenyl Group Contributes to Activities of Phytoestrogen 8-Prenylnaringenin in Enhancing Bone Formation and Inhibiting Bone Resorption In Vitro. *Endocrinology* **2013**, *154*, 1202–1214. [CrossRef]

21. Indran, I.R.; Liang, R.L.Z.; Min, T.E.; Yong, E.-L. Preclinical studies and clinical evaluation of compounds from the genus *Epimedium* for osteoporosis and bone health. *Pharmacol. Ther.* **2016**, *162*, 188–205. [CrossRef]

22. Kim, H.-K.; Woo, E.-R.; Lee, H.-W.; Park, H.-R.; Kim, H.-N.; Jung, Y.-K.; Choi, J.-Y.; Chae, S.-W.; Kim, H.-R.; Chae, H.-J. The Correlation of *Salvia miltiorrhiza* Extract–Induced Regulation of Osteoclastogenesis with the Amount of Components Tanshinone I, Tanshinone IIA, Cryptotanshinone, and Dihydrotanshinone. *Immunopharmacol. Immunotoxicol.* **2008**, *30*, 347–364. [CrossRef]

23. Cheng, L.; Zhou, S.; Zhao, Y.; Sun, Y.; Xu, Z.; Yuan, B.; Chen, X. Tanshinone IIA attenuates osteoclastogenesis in ovariecto-mized mice by inactivating NF-κB and Akt signaling pathways. *Am. J. Transl. Res.* **2018**, *10*, 1457–1468.

24. Cui, L.; Liu, Y.-Y.; Wu, T.; Ai, C.-M.; Chen, H.-Q. Osteogenic effects of D(+)β-3,4-dihydroxyphenyl lactic acid (salvianic acid A, SAA) on osteoblasts and bone marrow stromal cells of intact and prednisone-treated rats. *Acta Pharmacol. Sin.* **2009**, *30*, 321–332. [CrossRef] [PubMed]

25. Cui, L.; Li, T.; Liu, Y.; Zhou, L.; Li, P.; Xu, B.; Huang, L.; Chen, Y.; Liu, Y.; Tian, X.; et al. Salvianolic Acid B Prevents Bone Loss in Prednisone-Treated Rats through Stimulation of Osteogenesis and Bone Marrow Angiogenesis. *PLoS ONE* **2012**, *7*, e34647. [CrossRef]

26. Kim, M.-H.; Lim, H.-J.; Bak, S.G.; Park, E.-J.; Jang, H.-J.; Lee, S.; Lee, S.; Lee, K.M.; Cheong, S.H.; Lee, S.-J.; et al. Eudebeiolide B Inhibits Osteoclastogenesis and Prevents Ovariectomy-Induced Bone Loss by Regulating RANKL-Induced NF-κB, c-Fos and Calcium Signaling. *Pharmaceuticals* **2020**, *13*, 468. [CrossRef]

27. Scheiber, M.D.; Liu, J.H.; Subbiah, M.T.R.; Rebar, R.W.; Setchell, K.D.R. Dietary inclusion of whole soy foods results in significant reductions in clinical risk factors for osteoporosis and cardiovascular disease in normal postmenopausal women. *Menopause* **2001**, *8*, 384–392. [CrossRef]

28. Chiechi, L.M.; Secreto, G.; D'Amore, M.; Fanelli, M.; Venturelli, E.; Cantatore, F.; Valerio, T.; LaSelva, G.; Loizzi, P. Efficacy of a soy rich diet in preventing postmenopausal osteoporosis: The Menfis randomized trial. *Maturitas* **2002**, *42*, 295–300. [CrossRef]

29. Chen, Y.-M.; Ho, S.C.; Lam, S.S.H.; Ho, S.S.S.; Woo, J.L.F. Beneficial effect of soy isoflavones on bone mineral content was modified by years since menopause, body weight, and calcium intake: A double-blind, randomized, controlled trial. *Menopause* **2004**, *11*, 246–254. [CrossRef]

30. Ye, Y.-B.; Tang, X.-Y.; Verbruggen, M.A.; Su, Y.-X. Soy isoflavones attenuate bone loss in early postmenopausal Chinese women. *Eur. J. Nutr.* **2006**, *45*, 327–334. [CrossRef]

31. Wu, J.; Oka, J.; Ezaki, J.; Ohtomo, T.; Ueno, T.; Uchiyama, S.; Toda, T.; Uehara, M.; Ishimi, Y. Possible role of equol status in the effects of isoflavone on bone and fat mass in postmenopausal Japanese women. *Menopause* **2007**, *14*, 866–874. [CrossRef]

32. Shedd-Wise, K.M.; Alekel, D.L.; Hofmann, H.; Hanson, K.B.; Schiferl, D.J.; Hanson, L.N.; Van Loan, M.D. The Soy Isoflavones for Reducing Bone Loss Study: 3-Yr Effects on pQCT Bone Mineral Density and Strength Measures in Postmenopausal Women. *J. Clin. Densitom.* **2011**, *14*, 47–57. [CrossRef]

33. Lee, H.; Choue, R.; Lim, H. Effect of soy isoflavones supplement on climacteric symptoms, bone biomarkers, and quality of life in Korean postmenopausal women: A randomized clinical trial. *Nutr. Res. Pract.* **2017**, *11*, 223–231. [CrossRef] [PubMed]

34. Tit, D.M.; Bungau, S.; Iovan, C.; Cseppento, D.C.N.; Endres, L.; Sava, C.; Sabau, A.M.; Furau, G.; Furau, C. Effects of the Hormone Replacement Therapy and of Soy Isoflavones on Bone Resorption in Postmenopause. *J. Clin. Med.* **2018**, *7*, 297. [CrossRef]

35. Sathyapalan, T.; Aye, M.; Rigby, A.S.; Fraser, W.D.; Thatcher, N.J.; Kilpatrick, E.S.; Atkin, S.L. Soy Reduces Bone Turnover Markers in Women During Early Menopause: A Randomized Controlled Trial. *J. Bone Miner. Res.* **2016**, *32*, 157–164. [CrossRef] [PubMed]

36. Levis, S.; Strickman-Stein, N.; Ganjei-Azar, P.; Xu, P.; Doerge, D.R.; Krischer, J. Soy Isoflavones in the Prevention of Menopausal Bone Loss and Menopausal Symptoms. *Arch. Intern. Med.* **2011**, *171*, 1363–1369. [CrossRef]

37. Kreijkamp-Kaspers, S.; Kok, L.; Grobbee, D.E.; De Haan, E.H.F.; Aleman, A.; Lampe, J.W.; Van Der Schouw, Y.T. Effect of Soy Protein Containing Isoflavones on Cognitive Function, Bone Mineral Density, and Plasma Lipids in Postmenopausal Women. *JAMA* **2004**, *292*, 65–74. [CrossRef] [PubMed]

38. Brink, E.; Coxam, V.; Robins, S.; Wahala, K.; Cassidy, A.; Branca, F.; PHYTOS Investigators. Long-term consumption of isoflavone-enriched foods does not affect bone mineral density, bone metabolism, or hormonal status in early postmenopausal women: A randomized, double-blind, placebo controlled study. *Am. J. Clin. Nutr.* **2008**, *87*, 761–770. [CrossRef]

39. Marini, H.; Minutoli, L.; Polito, F.; Bitto, A.; Altavilla, D.; Atteritano, M.; Gaudio, A.; Mazzaferro, S.; Frisina, A.; Frisina, N.; et al. Effects of the Phytoestrogen Genistein on Bone Metabolism in Osteopenic Postmenopausal Women. *Ann. Intern. Med.* **2007**, *146*, 839–847. [CrossRef] [PubMed]

40. Marini, H.R.; Bitto, A.; Altavilla, D.; Burnett, B.P.; Polito, F.; Di Stefano, V.; Minutoli, L.; Atteritano, M.; Levy, R.M.; D'Anna, R.; et al. Breast Safety and Efficacy of Genistein Aglycone for Postmenopausal Bone Loss: A Follow-Up Study. *J. Clin. Endocrinol. Metab.* **2008**, *93*, 4787–4796. [CrossRef]

41. Atteritano, M.; Mazzaferro, S.; Frisina, A.; Cannata, M.L.; Bitto, A.; D'Anna, R.; Squadrito, F.; Macrì, I.; Frisina, N.; Buemi, M. Genistein effects on quantitative ultrasound parameters and bone mineral density in osteopenic postmenopausal women. *Osteoporos. Int.* **2009**, *20*, 1947–1954. [CrossRef]

42. Lappe, J.; Kunz, I.; Bendik, I.; Prudence, K.; Weber, P.; Recker, R.; Heaney, R.P. Effect of a combination of genistein, polyunsaturated fatty acids and vitamins D3 and K1 on bone mineral density in postmenopausal women: A randomized, placebo-controlled, double-blind pilot study. *Eur. J. Nutr.* **2013**, *52*, 203–215. [CrossRef]

43. Arcoraci, V.; Atteritano, M.; Squadrito, F.; D'Anna, R.; Marini, H.R.; Santoro, D.; Minutoli, L.; Messina, S.; Altavilla, D.; Bitto, A. Antiosteoporotic Activity of Genistein Aglycone in Postmenopausal Women: Evidence from a Post-Hoc Analysis of a Multicenter Randomized Controlled Trial. *Nutrients* **2017**, *9*, 179. [CrossRef]

44. Clifton-Bligh, P.B.; Baber, R.J.; Fulcher, G.R.; Nery, M.-L.; Moreton, T. The effect of isoflavones extracted from red clover (Rimostil) on lipid and bone metabolism. *Menopause* **2001**, *8*, 259–265. [CrossRef] [PubMed]

45. Atkinson, C.; Compston, J.E.; Day, N.E.; Dowsett, M.; Bingham, S.A. The effects of phytoestrogen isoflavones on bone density in women: A double-blind, randomized, placebo-controlled trial. *Am. J. Clin. Nutr.* **2004**, *79*, 326–333. [CrossRef] [PubMed]

46. Lambert, M.N.T.; Thybo, C.B.; Lykkeboe, S.; Rasmussen, L.M.; Frette, X.; Christensen, L.P.; Jeppesen, P.B. Combined bioavailable isoflavones and probiotics improve bone status and estrogen metabolism in postmenopausal osteopenic women: A randomized controlled trial. *Am. J. Clin. Nutr.* **2017**, *106*, ajcn153353-920. [CrossRef]

47. Thorup, A.C.; Lambert, M.N.; Kahr, H.S.; Bjerre, M.; Jeppesen, P.B. Intake of Novel Red Clover Supplementation for 12 Weeks Improves Bone Status in Healthy Menopausal Women. *Evid.-Based Complement. Altern. Med.* **2015**, *2015*, 9138. [CrossRef]

48. Powles, T.J.; Howell, A.; Evans, D.G.; McCloskey, E.V.; Ashley, S.; Greenhalgh, R.; Affen, J.; Flook, L.A.; Tidy, A. Red clover isoflavones are safe and well tolerated in women with a family history of breast cancer. *Menopause Int. Integr. J. Postreproductive Health* **2008**, *14*, 6–12. [CrossRef]

49. Schult, T.M.K.; Ensrud, K.E.; Blackwell, T.; Ettinger, B.; Wallace, R.; Tice, J.A. Effect of isoflavones on lipids and bone turnover markers in menopausal women. *Maturitas* **2004**, *48*, 209–218. [CrossRef] [PubMed]

50. Manonai, J.; Chittacharoen, A.; Udomsubpayakul, U.; Theppisai, H.; Theppisai, U. Effects and safety of Pueraria mirifica on lipid profiles and biochemical markers of bone turnover rates in healthy postmenopausal women. *Menopause* **2008**, *15*, 530–535. [CrossRef]

51. Okamura, S.; Sawada, Y.; Satoh, T.; Sakamoto, H.; Saito, Y.; Sumino, H.; Takizawa, T.; Kogure, T.; Chaichantipyuth, C.; Higuchi, Y.; et al. Pueraria Mirifica Phytoestrogens Improve Dyslipidemia in Postmenopausal Women Probably by Activating Estrogen Receptor Subtypes. *Tohoku J. Exp. Med.* **2008**, *216*, 341–351. [CrossRef]

52. Zhang, G.; Qin, L.; Shi, Y. Epimedium-Derived Phytoestrogen Flavonoids Exert Beneficial Effect on Preventing Bone Loss in Late Postmenopausal Women: A 24-Month Randomized, Double-Blind and Placebo-Controlled Trial. *J. Bone Miner. Res.* **2007**, *22*, 1072–1079. [CrossRef]

53. Arjmandi, B.H.; Khalil, D.A.; Lucas, E.A.; Georgis, A.; Stoecker, B.J.; Hardin, C.; Payton, M.E.; Wild, R.A. Dried Plums Improve Indices of Bone Formation in Postmenopausal Women. *J. Women's Heal. Gender-Based Med.* **2002**, *11*, 61–68. [CrossRef]

54. Hooshmand, S.; Brisco, J.R.Y.; Arjmandi, B.H. The effect of dried plum on serum levels of receptor activator of NF-κB ligand, osteoprotegerin and sclerostin in osteopenic postmenopausal women: A randomised controlled trial. *Br. J. Nutr.* **2014**, *112*, 55–60. [CrossRef] [PubMed]

55. Hooshmand, S.; Kern, M.; Metti, D.; Shamloufard, P.; Chai, S.C.; Johnson, S.A.; Payton, M.E.; Arjmandi, B.H. The effect of two doses of dried plum on bone density and bone biomarkers in osteopenic postmenopausal women: A randomized, controlled trial. *Osteoporos. Int.* **2016**, *27*, 2271–2279. [CrossRef]

56. Arjmandi, B.; George, K.; Ormsbee, L.; Akhavan, N.; Munoz, J.; Foley, E.; Siebert, S. The Short-Term Effects of Prunes in Preventing Inflammation and Improving Indices of Bone Health in Osteopenic Men. *Curr. Dev. Nutr.* **2020**, *4*, 5. [CrossRef]

57. Corletto, F. Female climacteric osteoporosis therapy with titrated horsetail (*Equisetum arvense*) extract plus calcium (osteosil calcium): Randomized double blind study. *Minerva Ortop. Traumatol.* **1999**, *50*, 201–206.

58. Wuttke, W.; Gorkow, C.; Seidlová-Wuttke, D. Effects of black cohosh (*Cimicifuga racemosa*) on bone turnover, vaginal mucosa, and various blood parameters in postmenopausal women. *Menopause* **2006**, *13*, 185–196. [CrossRef] [PubMed]

59. García-Pérez, M.A.; Pineda, B.; Hermenegildo, C.; Tarín, J.J.; Cano, A. Isopropanolic *Cimicifuga racemosa* is favorable on bone markers but neutral on an osteoblastic cell line. *Fertil. Steril.* **2009**, *91*, 1347–1350. [CrossRef] [PubMed]

60. Bebenek, M.; Kemmler, W.; von Stengel, S.; Engelke, K.; Kalender, W.A. Effect of exercise and *Cimicifuga racemosa* (CR BNO 1055) on bone mineral density, 10-year coronary heart disease risk, and menopausal complaints. *Menopause* **2010**, *17*, 791–800. [CrossRef] [PubMed]

61. Chinnappan, S.M.; George, A.; Evans, M.; Anthony, J. Efficacy of *Labisia pumila* and *Eurycoma longifolia* standardised extracts on hot flushes, quality of life, hormone and lipid profile of peri-menopausal and menopausal women: A randomised, placebo-controlled study. *Food Nutr. Res.* **2020**, *64*, 1–15. [CrossRef]

62. Křížová, L.; Dadáková, K.; Kašparovská, J.; Kašparovský, T. Isoflavones. *Molecules* **2019**, *24*, 1076. [CrossRef]

63. Cassidy, A.; Peñalvo, J.; Hollman, P. Bioavailability of isoflavones in humans. In *Flavonoids and Related Compounds: Bioavail-ability and Function*; CRC Press: Boca Raton, FL, USA, 2012; ISBN 9781439848272.

64. Jolly, J.J.; Chin, K.-Y.; Alias, E.; Chua, K.H.; Soelaiman, I.N. Protective Effects of Selected Botanical Agents on Bone. *Int. J. Environ. Res. Public Health* **2018**, *15*, 963. [CrossRef]

65. Collison, M.W. Determination of Total Soy Isoflavones in Dietary Supplements, Supplement Ingredients, and Soy Foods by High-Performance Liquid Chromatography with Ultraviolet Detection: Collaborative Study. *J. AOAC Int.* **2008**, *91*, 489–500. [CrossRef] [PubMed]

66. Zhang, X.; Shu, X.-O.; Li, H.; Yang, G.; Li, Q.; Gao, Y.-T.; Zheng, W. Prospective Cohort Study of Soy Food Consumption and Risk of Bone Fracture Among Postmenopausal Women. *Arch. Intern. Med.* **2005**, *165*, 1890–1895. [CrossRef] [PubMed]

67. Dennison, E.; Yoshimura, N.; Hashimoto, T.; Cooper, C. Bone loss in Great Britain and Japan: A comparative longitudinal study. *Bone* **1998**, *23*, 379–382. [CrossRef] [PubMed]

68. Ross, P.D.; Norimatsu, H.; Davis, J.W.; Yano, K.; Wasnich, R.D.; Fujiwara, S.; Hosoda, Y.; Melton, I.L.J. A Comparison of Hip Fracture Incidence among Native Japanese, Japanese Americans, and American Caucasians. *Am. J. Epidemiol.* **1991**, *133*, 801–809. [CrossRef] [PubMed]

69. Atteritano, M.; Pernice, F.; Mazzaferro, S.; Mantuano, S.; Frisina, A.; D'Anna, R.; Cannata, M.L.; Bitto, A.; Squadrito, F.; Frisina, N.; et al. Effects of phytoestrogen genistein on cytogenetic biomarkers in postmenopausal women: 1 year randomized, placebo-controlled study. *Eur. J. Pharmacol.* **2008**, *589*, 22–26. [CrossRef]

70. Pawlowski, J.W.; Martin, B.R.; McCabe, G.P.; McCabe, L.D.; Jackson, G.S.; Peacock, M.; Barnes, S.; Weaver, C.M. Impact of equol-producing capacity and soy-isoflavone profiles of supplements on bone calcium retention in postmenopausal women: A randomized crossover trial. *Am. J. Clin. Nutr.* **2015**, *102*, 695–703. [CrossRef]

71. Lambert, M.N.T.; Hu, L.M.; Jeppesen, P.B. A systematic review and meta-analysis of the effects of isoflavone formulations against estrogen-deficient bone resorption in peri- and postmenopausal women. *Am. J. Clin. Nutr.* **2017**, *106*, ajcn151464-811. [CrossRef]

72. Akhlaghi, M.; Nasab, M.G.; Riasatian, M.; Sadeghi, F. Soy isoflavones prevent bone resorption and loss, a systematic review and meta-analysis of randomized controlled trials. *Crit. Rev. Food Sci. Nutr.* **2019**, *60*, 2327–2341. [CrossRef]

73. Abdi, F.; Alimoradi, Z.; Haqi, P.; Mahdizad, F. Effects of phytoestrogens on bone mineral density during the menopause transition: A systematic review of randomized, controlled trials. *Climacteric* **2016**, *19*, 535–545. [CrossRef]

74. Liu, J.; Ho, S.C.; Su, Y.-X.; Chen, W.-Q.; Zhang, C.-X.; Chen, Y.-M. Effect of long-term intervention of soy isoflavones on bone mineral density in women: A meta-analysis of randomized controlled trials. *Bone* **2009**, *44*, 948–953. [CrossRef] [PubMed]

75. Jin, Y.-X.; Wu, P.; Mao, Y.-F.; Wang, B.; Zhang, J.-F.; Chen, W.-L.; Liu, Z.; Shi, X.-L. Chinese Herbal Medicine for Osteoporosis: A Meta-analysis of Randomized Controlled Trials. *J. Clin. Densitom.* **2017**, *20*, 516–525. [CrossRef]

76. Zhai, Y.; Li, Y.; Wang, Y.; Cui, J.; Feng, K.; Kong, X.; Chen, L. Psoralidin, a prenylated coumestan, as a novel anti-osteoporosis candidate to enhance bone formation of osteoblasts and decrease bone resorption of osteoclasts. *Eur. J. Pharmacol.* **2017**, *801*, 62–71. [CrossRef]

77. Chopra, B.; Dhingra, A.K.; Dhar, K.L. *Psoralea corylifolia* L. (Buguchi)—Folklore to modern evidence: Review. *Fitoterapia* **2013**, *90*, 44–56. [CrossRef]

78. Nieves, J.W. Skeletal effects of nutrients and nutraceuticals, beyond calcium and vitamin D. *Osteoporos. Int.* **2013**, *24*, 771–786. [CrossRef]

79. Kim, M.R.; Kim, H.J.; Yu, S.H.; Lee, B.S.; Jeon, S.Y.; Lee, J.J.; Lee, Y.C. Combination of Red Clover and Hops Extract Improved Menopause Symptoms in an Ovariectomized Rat Model. *Evid.-Based Complement. Altern. Med.* **2020**, *2020*, 1391. [CrossRef]

80. Geller, S.E.; Studee, L. Soy and red clover for mid-life and aging. *Climacteric* **2006**, *9*, 245–263. [CrossRef]

81. Booth, N.L.; Piersen, C.E.; Banuvar, S.; Geller, S.E.; Shulman, L.P.; Farnsworth, N.R. Clinical studies of red clover (*Trifolium pratense*) dietary supplements in menopause: A literature review. *Menopause* **2006**, *13*, 251–264. [CrossRef] [PubMed]

82. Park, Y.; Moon, H.-J.; Paik, D.-J.; Kim, D.-Y. Effect of dietary legumes on bone-specific gene expression in ovariectomized rats. *Nutr. Res. Pract.* **2013**, *7*, 185–191. [CrossRef] [PubMed]

83. Byun, J.S.; Han, Y.S.; Lee, S.S. The Effects of Yellow Soybean, Black Soybean, and Sword Bean on Lipid Levels and Oxidative Stress in Ovariectomized Rats. *Int. J. Vitam. Nutr. Res.* **2010**, *80*, 97–106. [CrossRef]

84. Hinton, P.S.; Ortinau, L.C.; Dirkes, R.K.; Shaw, E.L.; Richard, M.W.; Zidon, T.Z.; Britton, S.L.; Koch, L.G.; Vieira-Potter, V.J. Soy protein improves tibial whole-bone and tissue-level biomechanical properties in ovariectomized and ovary-intact, low-fit female rats. *Bone Rep.* **2018**, *8*, 244–254. [CrossRef]

85. Akao, M.; Abe, R.; Sato, N.; Hasegawa-Tanigome, A.; Kumagai, H.; Kumagai, H. Prevention of Osteoporosis by Oral Administration of Phytate-Removed and Deamidated Soybean β-Conglycinin. *Int. J. Mol. Sci.* **2015**, *16*, 2117–2129. [CrossRef]

86. Choi, C.-W.; Choi, S.-W.; Kim, H.-J.; Lee, K.-S.; Kim, S.-H.; Kim, S.-L.; Do, S.H.; Seo, W.-D. Germinated soy germ with increased soyasaponin Ab improves BMP-2-induced bone formation and protects against in vivo bone loss in osteoporosis. *Sci. Rep.* **2018**, *8*, 12970. [CrossRef]

87. Noh, D.; Lim, Y.; Lee, H.; Kim, H.; Kwon, O. Soybean-Hop Alleviates Estrogen Deficiency-Related Bone Loss and Metabolic Dysfunction in Ovariectomized Rats Fed a High-Fat Diet. *Molecules* **2018**, *23*, 1205. [CrossRef]

88. Kim, J.-S.; Lee, H.; Nirmala, F.S.; Jung, C.H.; Jang, Y.-J.; Ha, T.-Y.; Ahn, J. Dry-Fermented Soybean Food (Cheonggukjang) Ameliorates Senile Osteoporosis in the Senescence-Accelerated Mouse Prone 6 Model. *J. Med. Food* **2019**, *22*, 1047–1057. [CrossRef]

89. Zakłos-Szyda, M.; Budryn, G.; Grzelczyk, J.; Pérez-Sánchez, H.; Żyżelewicz, D. Evaluation of Isoflavones as Bone Resorption Inhibitors upon Interactions with Receptor Activator of Nuclear Factor-κB Ligand (RANKL). *Molecules* **2020**, *25*, 206. [CrossRef]

90. Chen, W.-F.; Wong, M.-S. Genistein modulates the effects of parathyroid hormone in human osteoblastic SaOS-2 cells. *Br. J. Nutr.* **2006**, *95*, 1039–1047. [CrossRef]

91. Jung, W.-K.; Choi, I.-W.; Hong, G.-E.; Pyun, C.-W.; Park, K.-K.; Seo, S.-K.; Lee, C.-H. Effects of Isoflavone Aglycone-rich Fermented Soybean Paste Extracts on Osteoblastic Differentiation of MG-63 Cells. *J. Korean Soc. Appl. Biol. Chem.* **2010**, *53*, 803–809. [CrossRef]

92. Ho, M.-X.; Poon, C.C.-W.; Wong, K.-C.; Qiu, Z.-C.; Wong, M.-S. Icariin, but Not Genistein, Exerts Osteogenic and Anti-apoptotic Effects in Osteoblastic Cells by Selective Activation of Non-genomic ERα Signaling. *Front. Pharmacol.* **2018**, *9*, 474. [CrossRef]

93. Thangavel, P.; Puga-Olguín, A.; Rodríguez-Landa, J.F.; Zepeda, R.C. Genistein as Potential Therapeutic Candidate for Menopausal Symptoms and Other Related Diseases. *Molecules* **2019**, *24*, 3892. [CrossRef]

94. Chen, C.; Zheng, H.; Qi, S. Genistein and Silicon Synergistically Protects Against Ovariectomy-Induced Bone Loss Through Upregulating OPG/RANKL Ratio. *Biol. Trace Element Res.* **2018**, *188*, 441–450. [CrossRef]

95. Qi, S. Synergistic Effects of Genistein and Zinc on Bone Metabolism and the Femoral Metaphyseal Histomorphology in the Ovariectomized Rats. *Biol. Trace Elem. Res.* **2017**, *183*, 288–295. [CrossRef]

96. Hu, B.; Yu, B.; Tang, D.; Li, S.; Wu, Y. Daidzein promotes osteoblast proliferation and differentiation in OCT1 cells through stimulating the activation of BMP-2/Smads pathway. *Genet. Mol. Res.* **2016**, *15*. [CrossRef]

97. Jin, X.; Sun, J.; Yu, B.; Wang, Y.; Sun, W.J.; Yang, J.; Huang, S.H.; Xie, W.L. Daidzein stimulates osteogenesis facilitating proliferation, differentiation, and antiapoptosis in human osteoblast-like MG-63 cells via estrogen receptor–dependent MEK/ERK and PI3K/Akt activation. *Nutr. Res.* **2017**, *42*, 20–30. [CrossRef] [PubMed]

98. Wang, S.; Zhang, S.; Wang, S.; Gao, P.; Dai, L. A comprehensive review on Pueraria: Insights on its chemistry and medicinal value. *Biomed. Pharmacother.* **2020**, *131*, 110734. [CrossRef]

99. Li, B.; Liu, M.; Wang, Y.; Gong, S.; Yao, W.; Li, W.; Gao, H.; Wei, M. Puerarin improves the bone micro-environment to inhibit OVX-induced osteoporosis via modulating SCFAs released by the gut microbiota and repairing intestinal mucosal integrity. *Biomed. Pharmacother.* **2020**, *132*, 110923. [CrossRef] [PubMed]

100. Zhang, G.; Wang, Y.; Tang, G.; Ma, Y. Puerarin inhibits the osteoclastogenesis by inhibiting RANKL-dependent and –independent autophagic responses. *BMC Complement. Altern. Med.* **2019**, *19*, 269. [CrossRef]

101. Zhang, Y.; Zeng, X.; Zhang, L.; Zheng, X. Stimulatory Effect of Puerarin on Bone Formation through Activation of PI3K/Akt Pathway in Rat Calvaria Osteoblasts. *Planta Medica* **2007**, *73*, 341–347. [CrossRef]

102. Shan, Z.; Cheng, N.; Huang, R.; Zhao, B.; Zhou, Y. Puerarin promotes the proliferation and differentiation of MC3T3-E1 cells via microRNA-106b by targeting receptor activator of nuclear factor-κB ligand. *Exp. Ther. Med.* **2017**, *15*, 55–60. [CrossRef]

103. Ma, H.; He, X.; Yang, Y.; Li, M.; Hao, D.; Jia, Z. The genus Epimedium: An ethnopharmacological and phytochemical review. *J. Ethnopharmacol.* **2011**, *134*, 519–541. [CrossRef]

104. Jia, M.; Nie, Y.; Cao, D.-P.; Xue, Y.-Y.; Wang, J.-S.; Zhao, L.; Rahman, K.; Zhang, Q.-Y.; Qin, L.-P. Potential Antiosteoporotic Agents from Plants: A Comprehensive Review. *Evid.-Based Complement. Altern. Med.* **2012**, *2012*, 364604. [CrossRef]

105. Zhao, B.-J.; Wang, J.; Song, J.; Gu, J.-F.; Yuan, J.-R.; Zhang, L.; Jiang, J.; Feng, L.; Jia, X.-B. Beneficial Effects of a Flavonoid Fraction of Herba Epimedii on Bone Metabolism in Ovariectomized Rats. *Planta Med.* **2016**, *82*, 322–329. [CrossRef]

106. Xu, H.; Zhou, S.; Qu, R.; Yang, Y.; Gong, X.; Hong, Y.; Jin, A.; Huang, X.; Dai, Q.; Jiang, L. Icariin prevents oestrogen deficiency–induced alveolar bone loss through promoting osteogenesis via STAT3. *Cell Prolif.* **2020**, *53*, e12743. [CrossRef]

107. Keiler, A.; Zierau, O.; Kretzschmar, G. Hop Extracts and Hop Substances in Treatment of Menopausal Complaints. *Planta Med.* **2013**, *79*, 576–579. [CrossRef] [PubMed]

108. Milligan, S.R.; Kalita, J.C.; Pocock, V.; Van De Kauter, V.; Stevens, J.F.; Deinzer, M.L.; Rong, H.; De Keukeleire, D. The Endocrine Activities of 8-Prenylnaringenin and Related Hop (*Humulus lupulus* L.) Flavonoids. *J. Clin. Endocrinol. Metab.* **2000**, *85*, 4912–4915. [CrossRef]

109. Ban, Y.-H.; Yon, J.-M.; Cha, Y.; Choi, J.; An, E.S.; Guo, H.; Seo, D.W.; Kim, T.-S.; Lee, S.-P.; Kim, J.-C.; et al. A Hop Extract Lifenol® Improves Postmenopausal Overweight, Osteoporosis, and Hot Flash in Ovariectomized Rats. *Evid.-Based Complement. Altern. Med.* **2018**, *2018*, 2929107. [CrossRef]

110. Keiler, A.M.; Helle, J.; Bader, M.I.; Ehrhardt, T.; Nestler, K.; Kretzschmar, G.; Bernhardt, R.; Vollmer, G.; Nikolić, D.; Bolton, J.L.; et al. A standardized *Humulus lupulus* (L.) ethanol extract partially prevents ovariectomy-induced bone loss in the rat without induction of adverse effects in the uterus. *Phytomedicine* **2017**, *34*, 50–58. [CrossRef]

111. Li, J.; Zeng, L.; Xie, J.; Yue, Z.; Deng, H.; Ma, X.; Zheng, C.; Wu, X.; Luo, J.; Liu, M. Inhibition of Osteoclastogenesis and Bone Resorption in vitro and in vivo by a prenylflavonoid xanthohumol from hops. *Sci. Rep.* **2015**, *5*, 17605. [CrossRef] [PubMed]

112. Jeong, H.M.; Han, E.H.; Jin, Y.H.; Choi, Y.H.; Lee, K.Y.; Jeong, H.G. Xanthohumol from the hop plant stimulates osteoblast differentiation by RUNX2 activation. *Biochem. Biophys. Res. Commun.* **2011**, *409*, 82–89. [CrossRef]

113. Luo, D.; Kang, L.; Ma, Y.; Chen, H.; Kuang, H.; Huang, Q.; He, M.; Peng, W. Effects and mechanisms of 8-prenylnaringenin on osteoblast MC 3T3-E1 and osteoclast-like cells RAW 264.7. *Food Sci. Nutr.* **2014**, *2*, 341–350. [CrossRef]

114. Arjmandi, B.H.; Johnson, S.A.; Pourafshar, S.; Navaei, N.; George, K.S.; Hooshmand, S.; Chai, S.C.; Akhavan, N.S. Bone-Protective Effects of Dried Plum in Postmenopausal Women: Efficacy and Possible Mechanisms. *Nutrients* **2017**, *9*, 496. [CrossRef]

115. Hooshmand, S.; Chai, S.C.; Saadat, R.L.; Payton, M.E.; Brummel-Smith, K.; Arjmandi, B.H. Comparative effects of dried plum and dried apple on bone in postmenopausal women. *Br. J. Nutr.* **2011**, *106*, 923–930. [CrossRef]

116. Simonavice, E.; Liu, P.-Y.; Ilich, J.Z.; Kim, J.-S.; Arjmandi, B.; Panton, L.B. The effects of a 6-month resistance training and dried plum consumption intervention on strength, body composition, blood markers of bone turnover, and inflammation in breast cancer survivors. *Appl. Physiol. Nutr. Metab.* **2014**, *39*, 730–739. [CrossRef] [PubMed]

117. Rajput, R.; Wairkar, S.; Gaud, R. Nutraceuticals for better management of osteoporosis: An overview. *J. Funct. Foods* **2018**, *47*, 480–490. [CrossRef]

118. Arjmandi, B.H.; Johnson, C.D.; Campbell, S.C.; Hooshmand, S.; Chai, S.C.; Akhter, M.P. Combining Fructooligosaccharide and Dried Plum Has the Greatest Effect on Restoring Bone Mineral Density Among Select Functional Foods and Bioactive Compounds. *J. Med. Food* **2010**, *13*, 312–319. [CrossRef]

119. Bu, S.Y.; Lucas, E.A.; Franklin, M.; Marlow, D.; Brackett, D.J.; Boldrin, E.A.; Devareddy, L.; Arjmandi, B.H.; Smith, B.J. Comparison of dried plum supplementation and intermittent PTH in restoring bone in osteopenic orchidectomized rats. *Osteoporos. Int.* **2007**, *18*, 931–942. [CrossRef] [PubMed]

120. Deyhim, F.; Stoecker, B.J.; Brusewitz, G.H.; Devareddy, L.; Arjmandi, B.H. Dried plum reverses bone loss in an osteopenic rat model of osteoporosis. *Menopause* **2005**, *12*, 755–762. [CrossRef]

121. Rendina, E.; Hembree, K.D.; Davis, M.R.; Marlow, D.; Clarke, S.L.; Halloran, B.P.; Lucas, E.A.; Smith, B.J. Dried Plum's Unique Capacity to Reverse Bone Loss and Alter Bone Metabolism in Postmenopausal Osteoporosis Model. *PLoS ONE* **2013**, *8*, e60569. [CrossRef]

122. Graef, J.L.; Ouyang, P.; Wang, Y.; Rendina-Ruedy, E.; Lerner, M.R.; Marlow, D.; Lucas, E.A.; Smith, B.J. Dried plum polyphenolic extract combined with vitamin K and potassium restores trabecular and cortical bone in osteopenic model of postmenopausal bone loss. *J. Funct. Foods* **2018**, *42*, 262–270. [CrossRef] [PubMed]

123. Bu, S.Y.; Lerner, M.; Stoecker, B.J.; Boldrin, E.; Brackett, D.J.; Lucas, E.A.; Smith, B.J. Dried Plum Polyphenols Inhibit Osteoclastogenesis by Downregulating NFATc1 and Inflammatory Mediators. *Calcif. Tissue Int.* **2008**, *82*, 475–488. [CrossRef]

124. Bu, S.Y.; Hunt, T.S.; Smith, B.J. Dried plum polyphenols attenuate the detrimental effects of TNF-α on osteoblast function coincident with up-regulation of Runx2, Osterix and IGF-I. *J. Nutr. Biochem.* **2009**, *20*, 35–44. [CrossRef]

125. Pereira, C.B.; Gomes, P.S.; Costa-Rodrigues, J.; Palmas, R.A.; Vieira, L.; Ferraz, M.P.; Lopes, M.A.; Fernandes, M.H. *Equisetum arvense* hydromethanolic extracts in bone tissue regeneration: In vitro osteoblastic modulation and antibacterial activity. *Cell Prolif.* **2012**, *45*, 386–396. [CrossRef] [PubMed]

126. Costa-Rodrigues, J.; Carmo, S.C.; Silva, J.C.; Fernandes, M.H. Inhibition of human in vitro osteoclastogenesis by *Equisetum arvense*. *Cell Prolif.* **2012**, *45*, 566–576. [CrossRef]

127. Kotwal, S.D.; Badole, S.R. Anabolic therapy with *Equisetum arvense* along with bone mineralising nutrients in ovariectomized rat model of osteoporosis. *Indian J. Pharmacol.* **2016**, *48*, 312–315. [CrossRef] [PubMed]

128. Arbabzadegan, N.; Moghadamnia, A.A.; Kazemi, S.; Nozari, F.; Moudi, E.; Haghanifar, S. Effect of *Equisetum arvense* extract on bone mineral density in Wistar rats via digital radiography. *Casp. J. Intern. Med* **2019**, *10*, 176–182.

129. EFSA Panel on Dietetic Products, Nutrition and Allergies (NDA). Scientific opinion on the substantiation of health claims related to *Equisetum arvense* L. and invigoration of the body (ID 2437), maintenance of skin (ID 2438), maintenance of hair (ID 2438), maintenance of bone (ID 2439), and maintenance or achievement of a normal body weight (ID 2783) pursuant to Article 13 of Regulation (EC) No 1924/2006. *EFSA J.* **2009**, *7*, 1289.

130. Zepelin, H.-H.H.-V. 60 years of *Cimicifuga racemosa* medicinal products. *Wien. Med. Wochenschr.* **2017**, *167*, 147–159. [CrossRef]

131. Dragos, D.; Gilca, M.; Gaman, L.; Stoian, I.; Lupescu, O. Osteoprotective medicinal plants-part 1 (A human clinical evidence-based review). *Afr. J. Tradit. Complement. Altern. Med.* **2017**, *14*, 102–119. [CrossRef]

132. Carlisle, A.P.; Jessup, J.V.; Simpson, S.; Yoon, S. Effect of black cohosh on biochemical markers of bone remodeling in post-menopausal women. *J. Midwifery Women's Health* **2009**, *54*, 424. [CrossRef]

133. Ahn, B.-S.; Yang, M.; Jang, H.; Lee, H.J.; Moon, C.; Kim, J.-C.; Jung, U.; Jo, S.K.; Jang, J.-S.; Kim, S.-H. Evaluation of the Antiosteoporotic Potential of *Cimicifuga heracleifolia* in Female Mice. *Phytother. Res.* **2011**, *26*, 663–668. [CrossRef]

34. Li, J.; Liu, J.; He, C.; Yu, Z.; Du, Y.; Kadota, S.; Seto, H. Triterpenoids from *Cimicifugae rhizoma*, a novel class of inhibitors on bone resorption and ovariectomy-induced bone loss. *Maturitas* **2007**, *58*, 59–69. [CrossRef]
35. Choi, E.M. Deoxyactein stimulates osteoblast function and inhibits bone-resorbing mediators in MC3T3-E1 cells. *J. Appl. Toxicol.* **2011**, *33*, 190–195. [CrossRef]
36. Lee, Y.S.; Choi, E.M. Actein Isolated from Black Cohosh Promotes the Function of Osteoblastic MC3T3-E1 Cells. *J. Med. Food* **2014**, *17*, 414–423. [CrossRef]
37. Guo, Y.; Li, Y.; Xue, L.; Severino, R.P.; Gao, S.; Niu, J.; Qin, L.-P.; Zhang, D.; Brömme, D. *Salvia miltiorrhiza*: An ancient Chinese herbal medicine as a source for anti-osteoporotic drugs. *J. Ethnopharmacol.* **2014**, *155*, 1401–1416. [CrossRef]
38. Dragos, D.; Gilca, M.; Gaman, L.; Vlad, A.; Iosif, L.; Stoian, I.; Lupescu, O. Phytomedicine in Joint Disorders. *Nutrients* **2017**, *9*, 70. [CrossRef]
39. Kim, M.-H.; Jung, K.; Nam, K.-H.; Jang, H.-J.; Lee, S.W.; Kim, Y.; Park, C.S.; Lee, T.-H.; Park, J.H.; Choi, J.H.; et al. *Salvia plebeia* R.Br. inhibits signal transduction of IL-6 and prevents ovariectomy-induced bone loss by suppressing osteoclastogenesis. *Arch. Pharmacal Res.* **2016**, *39*, 1671–1681. [CrossRef]
40. Liu, H.; Zhu, R.; Wang, L.; Liu, C.; Ma, R.; Qi, B.; Chen, B.; Li, L.; Guo, Y.; Shi, S.; et al. Radix *Salviae miltiorrhizae* improves bone microstructure and strength through Wnt/β-catenin and osteoprotegerin/receptor activator for nuclear factor-κB ligand/cathepsin K signaling in ovariectomized rats. *Phytother. Res.* **2018**, *32*, 2487–2500. [CrossRef]
41. Gupta, T.; Das, N.; Imran, S. The Prevention and Therapy of Osteoporosis: A Review on Emerging Trends from Hormonal Therapy to Synthetic Drugs to Plant-Based Bioactives. *J. Diet. Suppl.* **2018**, *16*, 699–713. [CrossRef]
142. Gan, D.; Xu, X.; Chen, D.; Feng, P.; Xu, Z. Network Pharmacology-Based Pharmacological Mechanism of the Chinese Medicine *Rhizoma drynariae* Against Osteoporosis. *Med. Sci. Monit.* **2019**, *25*, 5700–5716. [CrossRef] [PubMed]
143. Jang, S.-A.; Hwang, Y.-H.; Kim, T.; Lee, A.; Ha, H. Anti-Osteoporotic and Anti-Adipogenic Effects of the Water Extract of *Drynaria roosii* Nakaike in Ovariectomized Mice Fed a High-Fat Diet. *Molecules* **2019**, *24*, 3051. [CrossRef]
144. Sun, X.; Wei, B.; Peng, Z.; Chen, X.; Fu, Q.; Wang, C.; Zhen, J.; Sun, J. A polysaccharide from the dried rhizome of *Drynaria fortunei* (Kunze) J. Sm. prevents ovariectomized (OVX)-induced osteoporosis in rats. *J. Cell. Mol. Med.* **2020**, *24*, 3692–3700. [CrossRef]
145. Liu, Y.; Wang, C.; Wang, G.; Sun, Y.; Deng, Z.; Chen, L.; Chen, K.; Tickner, J.; Kenny, J.; Song, D.; et al. Loureirin B suppresses RANKL-induced osteoclastogenesis and ovariectomized osteoporosis via attenuating NFATc1 and ROS activities. *Theranostics* **2019**, *9*, 4648–4662. [CrossRef]
146. Zou, B.; Zheng, J.; Deng, W.; Tan, Y.; Jie, L.; Qu, Y.; Yang, Q.; Ke, M.; Ding, Z.; Chen, Y.; et al. Kirenol inhibits RANKL-induced osteoclastogenesis and prevents ovariectomized-induced osteoporosis via suppressing the Ca2+-NFATc1 and Cav-1 signaling pathways. *Phytomedicine* **2021**, *80*, 153377. [CrossRef]
147. Nowak, B.; Matuszewska, A.; Nikodem, A.; Filipiak, J.; Landwójtowicz, M.; Sadanowicz, E.; Jędrzejuk, D.; Rzeszutko, M.; Zduniak, K.; Piasecki, T.; et al. Oral administration of kaempferol inhibits bone loss in rat model of ovariectomy-induced osteopenia. *Pharmacol. Rep.* **2017**, *69*, 1113–1119. [CrossRef]
148. Wong, S.K.; Chin, K.-Y.; Ima-Nirwana, S. The Osteoprotective Effects of Kaempferol: The Evidence From In Vivo And In Vitro Studies. *Drug Des. Dev. Ther.* **2019**, *13*, 3497–3514. [CrossRef]
149. Nowak, B.; Matuszewska, A.; Szandruk, M.; Matkowski, A.; Woźniak, D.; Zduniak, K.; Rzeszutko, M.; Landwójtowicz, M.; Jędrzejuk, D.; Piasecki, T.; et al. Effect of long-term administration of mangiferin from *Belamcanda chinensis* on bone metabolism in ovariectomized rats. *J. Funct. Foods* **2018**, *46*, 12–18. [CrossRef]
150. Ding, L.-Z.; Teng, X.; Zhang, Z.-B.; Zheng, C.-J.; Chen, S.-H. Mangiferin inhibits apoptosis and oxidative stress via BMP2/Smad-1 signaling in dexamethasone-induced MC3T3-E1 cells. *Int. J. Mol. Med.* **2018**, *41*, 2517–2526. [CrossRef]
151. Sekiguchi, Y.; Mano, H.; Nakatani, S.; Shimizu, J.; Kataoka, A.; Ogura, K.; Kimira, Y.; Ebata, M.; Wada, M. Mangiferin positively regulates osteoblast differentiation and suppresses osteoclast differentiation. *Mol. Med. Rep.* **2017**, *16*, 1328–1332. [CrossRef]
152. Wong, S.K.; Chin, K.-Y.; Ima-Nirwana, S. Quercetin as an Agent for Protecting the Bone: A Review of the Current Evidence. *Int. J. Mol. Sci.* **2020**, *21*, 6448. [CrossRef]

MDPI
St. Alban-Anlage 66
4052 Basel
Switzerland
Tel. +41 61 683 77 34
Fax +41 61 302 89 18
www.mdpi.com

Nutrients Editorial Office
E-mail: nutrients@mdpi.com
www.mdpi.com/journal/nutrients